ARCHITECTURAL
MODELMAKING

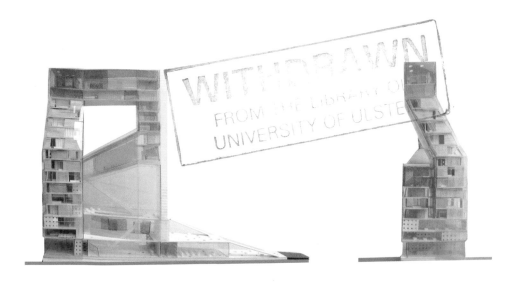

Published in 2010
by Laurence King Publishing Ltd
361–373 City Road
London EC1V 1LR
Tel +44 20 7841 6900
Fax +44 20 7841 6910
E enquiries@laurenceking.com
www.laurenceking.com

A catalogue record for this book is available from the British Library

ISBN 978 185669 670 8
Designed by John Round Design
Printed in China

NICK DUNN

ARCHITECTURAL MODELMAKING

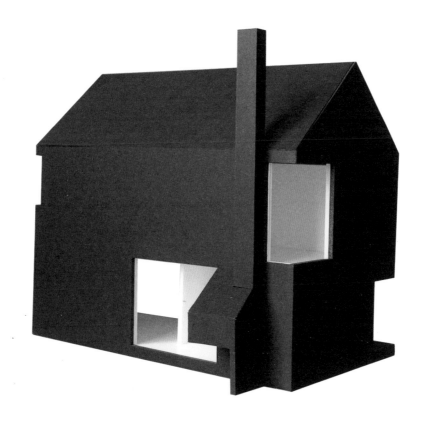

Laurence King Publishing

Contents

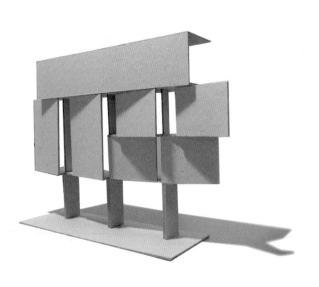

Related study material is available on the Laurence King website at
www.laurenceking.com

Introduction

Why we make models

The representation of creative ideas is of primary importance within any design-based discipline, and is particularly relevant in architecture where we often do not get to see the finished results, i.e. the building, until the very end of the design process. Initial concepts are developed through a process that enables the designer to investigate, revise and further refine ideas in increasing detail until such a point that the project's design is sufficiently consolidated to be constructed. Models can be extraordinarily versatile objects within this process, enabling designers to express thoughts creatively. Architects make models as a means of exploring and presenting the conception and development of ideas in three dimensions. One of the significant advantages of physical models is their immediacy, as they can communicate ideas about material, shape, size and colour in a highly accessible manner. The size of a model is often partially determined by the scale required at various stages of the design process, since models can illustrate a design project in relation to a city context, a landscape, as a remodelling or addition to an existing building or can even be constructed as full-size versions, typically referred to as 'prototypes'.

Throughout history, different types of models have been used extensively to explain deficiencies in knowledge. This is because models can be very provocative and evoke easy understanding as a method of communication. Our perception provides instant access to any part of a model, and to detailed as well as overall views. Familiar features can be quickly recognized, and this provides several ways for designers to draw attention to specific parts of a model. A significant advantage of using models is that they are a potentially rich source of information – providing three dimensions within which to present information, and the

Members of the Office for Metropolitan Architecture (OMA), with a design development model for the Universal Headquarters, Los Angeles.

opportunity to use a host of properties borrowed from the 'real' world, for example: size, shape, colour and texture. Therefore, since the 'language' of the model is so dense the 'encoding' of each piece of information can be more compact, with a resultant decrease in the decoding time in our understanding of it.

In order to understand architecture, it is critical to engage in a direct experience of space. As Tom Porter explains in *The Architect's Eye*, this is 'because architecture is concerned with the physical articulation of space; the amount and shape of the void contained and generated by buildings being as material a part of its existence as the substance of its fabric'.[1] The organization and representation of space is not the sole domain of architecture – other visually-orientated disciplines such as painting and sculpture are equally engaged with these tasks, but on different terms. The principal difference between these disciplines lies in a concern with the function of the final 'product'. In the case of painting or sculpture, such purposes are typically visual alone. By contrast, the creative process in architectural design often results in a building that has a responsibility to address additional concerns such as inhabitation, climatic considerations and maintenance. The considerable amounts of expense, resources and time invested in building architecture at its full-size scale demands that we need to be able to repeatedly describe, explore, predict and evaluate different properties of the design at various stages prior to construction. This raises an important issue concerning

modelmaking since, as with other modes of representation in architecture, it is not a 'neutral' means for the conveyance of ideas but is in fact the medium and mechanism through which concepts and designs are developed. This point is reinforced by Stan Allen writing about architectural drawing, as he suggests it is 'in some basic way impure, and unclassifiable. Its link to the reality it designates is complex and changeable. Like traditional painting and sculpture, it carries a mimetic trace, a representational shadow, which is transposed (spatially, across scale), into the built artefact. Drawings are, to some degree, scaled-down pictures of buildings. But to think of drawings as pictures cannot account either for the instrumentality of architectural representation nor for its capacity to render abstract ideas concrete.'[2] Considered in this manner, the discrimination on behalf of the modelmaker to decide which information to include, and therefore which to deliberately omit, to best represent design ideas becomes crucial.

As practitioners, architects are expected to have a highly evolved set of design skills, a core element of which is their ability to communicate their ideas using a variety of media. For the student learning architecture, the problem of communicating effectively so that the tutor may understand his or her design is central to the nature of design education; spatial ideas can become so elaborate that they have to be represented in some tangible form so that they can be described, explored and evaluated. However, this visual delivery of ideas is not simply for the benefit of a tutor or

Left
The team at the in-house modelmaking facility at Rogers Stirk Harbour + Partners works on a 1:200 model of The Leadenhall Building, London.

Right
Antoine Predock working on a clay model for Ohio State University's Recreation and Physical Activity Center, en route to a project presentation, 2001. On his wesbite, Predock refers to the importance his clay models have within his design process: 'compared to a drawing on paper, the models are very real; they are the building'.

review audience. The modelling of an architectural design has important advantages for a student during the design process as he attempts to express his ideas and translate them, allowing the designer to develop the initial concept. However, it is important to emphasise the performative nature of models at this point since they, as with other media and types of representation, are highly generative in terms of designing and are not simply used to transpose ideas. This establishes a continuous dialogue between design ideas and the method of representation, which flows until the process reaches a point of consolidation.

Fundamentally, the physical architectural model allows us to perceive the three-dimensional experience rather than having to try to imagine it. This not only enables a more effective method of communication to the receiver – such as the tutor, the client or the public – but also allows the transmitter – such as a student or architect – to develop and further revise the design. As Rolf Janke writes in the classic *Architectural Models*, 'the significance of a model lies not only in enabling (the architect) to depict in plastic terms the end-product of his deliberations, but in giving him the means – during the design process – of actually seeing and therefore controlling spatial problems'[3]; while Criss Mills asserts that 'models are capable of generating information in an amount of time comparable to that needed for drawing, and they offer one of the strongest exploration methods available'.[4]

At this stage it is worth expanding further upon the use of a model by a student of architecture or urban design, since it is a common assumption that an architect has sufficient experience and skills to employ a variety of design processes and methods of communication as required by the task in hand. When developing design ideas as a response to studio-project briefs, the employment of various methods of communication is a prerequisite for the thinking process necessary for a student to deal with the complexity of architectural design. Unlike the final presentation type frequently found in architectural practice, in educational environments models and drawings are not seen as end products in order to 'sell' the solution, but as vehicles for thought or tools with which ideas can be developed and expressed. More specifically, the use of different communication methods encourages greater exploration of a student's ideas. This is because different

1:20 model of a temporary pavilion by 6a architects in the process of being made. The intricate pattern directly replicates that of the final intervention and uses the same process of manufacture, albeit on a smaller scale.

A modelmaker at Alsop Architects in the process of assembling a presentation model for the CPlex project, West Bromwich (the completed building is known as The Public).

visualization methods and techniques provoke different thought processes, and inspire greater insight during the design process. Every model has a specific purpose and user, as it is not possible to embody all potential design ideas within a single one. In the first instance it may function purely as a design tool, allowing the designer to explore a particular idea or analyze successive developments. Secondly, it may be used to present or demonstrate design ideas to an audience – allowing others to share the designer's vision. Whilst it is tempting to classify different types of model, and indeed this book will look at the full spectrum available, it is apparent that most models are dynamic and have at least a dual function depending on who is using them and why and when they are employed throughout the design process. This short introduction seeks to demonstrate the significance of models not only as aids in the decision-making process, but also as a means of generating, searching and investigating creative impulses. Before moving onto the main part of the book there follows a brief overview of the role of the model in history, an explanation of the format of the book, and some basic information on the type of equipment necessary to begin making models.

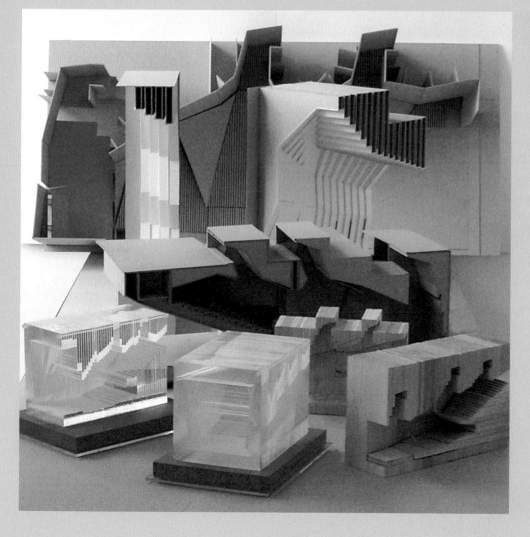

A variety of explorative models by Grafton Architects for their Università Luigi Bocconi scheme in Milan. Note the range of materials used and their resultant differing effects on the spaces within the models.

A quickly produced design-development model
made by UNStudio during their process of
composing the geometry for the Mercedes-Benz
Museum in Stuttgart. Models such as these
provide architects with flexible tools, through
which they can explore ideas in a fast and
effective manner.

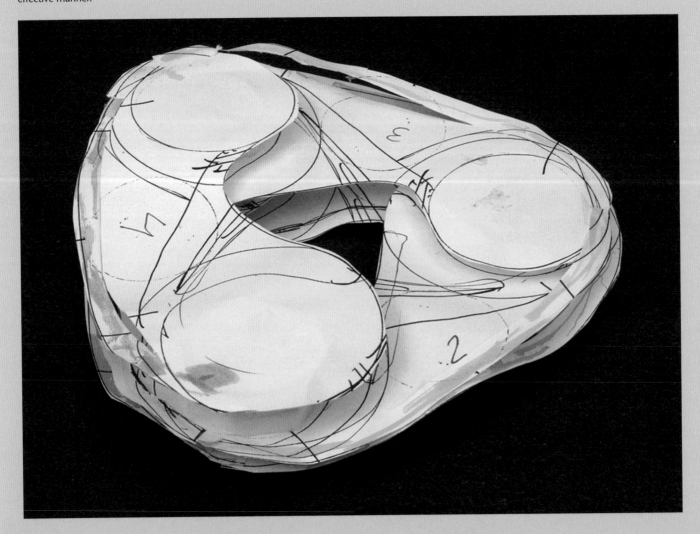

Below
The number of models made to investigate design ideas obviously varies from project to project, but a series of models that clearly communicate the process through which a concept has evolved are not uncommon, as shown here.

Above
Series of design-development models for Coop Himmelb(l)au's SEG Apartment Tower, Vienna, illustrating the increasing articulation of the tower's form in order to maximize its passive-energy performance.

Above
Sectional model for CPlex project
by Alsop Architects, investigating
internal characteristics of
the design.

Left
Competition model for the
Fourth Grace, Liverpool, by Rogers
Stirk Harbour + Partners, 1:500,
illustrating the urban scale of
the scheme.

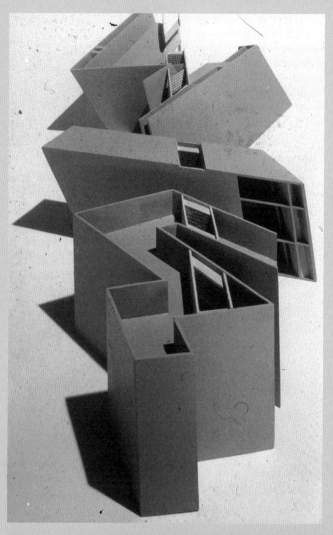

Left
Explorative model for Daniel Libeskind's design for the Jewish Museum, Berlin.
This model was made to examine the relationships between the voids of the building and its powerful generative geometry.

Below
Coop Himmelb(l)au's presentation model for Museum of Knowledge, Lyon, France (under construction). Note the careful use of lighting within the model to enhance the effect of the project's sculptural forms and their interplay with space.

A brief history

The first recorded use of architectural models dates back to the fifth century BC, when Herodotus, in Book V, Terpsichore, makes reference to a model of a temple. Whilst it may be inspiring to believe that scale models were used in the design of buildings from ancient civilizations, this appears highly unlikely. This is because the inaccuracies in translating scales at this time would have resulted in significant errors, but also because designs in antiquity were in fact developed with respect to cosmic measurements and proportions. Despite this, however, the production of repetitive architectural elements in large quantities was common, and the use of a full-size physical prototype as a three-dimensional template for the accurate replication of components such as column capitals was typical.

Architectural design continued in a similar manner through to the Middle Ages. Medieval architects travelled frequently to study and record vital proportions of Classical examples that would then be adapted in relation to a client's desires. Although models were not prevalent at this time, they would occasionally be constructed to scale from wood in order to enable detailed description to the client as well as to estimate materials and the cost of construction. This was largely because two-dimensional techniques of representation were comparatively under-developed. Therefore, despite the very early recording of an architectural model, there is no substantial evidence to suggest that such models were used again until this point: 'only since the fourteenth century has this form of representation become relevant to the practice of building; we know that a model of the Cathedral of Florence was made towards the end of the fourteenth century.'[5]

There appears to be an explanation for this emergence of the scale model as a method of design and communication. Unlike his predecessor and Gothic counterparts, the Renaissance architect had no equivalent frame of reference as he derived his style from fragments of Graeco-Roman architecture. The only method of checking the feasibility of these new architectural concepts was to build exploratory models. These models were even, when necessary, made using the actual materials proposed for the building itself.

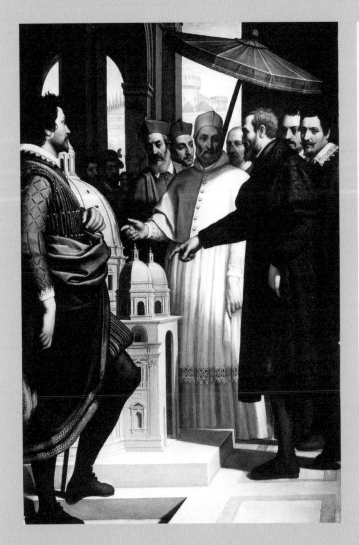

Therefore, from the early Renaissance on, an increasing number of architectural scale models exist, illustrating not only buildings but also urban districts and fortifications. Well-documented architectural models included those of the church of St Maclou in Rouen from the fifteenth century, the church of 'Schöne Maria' in Regensburg of 1520 and the pilgrimage church of Vierzehnheiligen by Balthasar Neumann, circa 1744. Such scale models were large prefabrications constructed in wood, plaster and clay. In contrast to the primitive structural models of the Middle Ages, these new models were expensive and extravagant – frequently including pull-away sections and detachable roofs and floors, both to allow internal viewing and to aid the development of the design.

During this period, the proliferation and status of the architectural scale model grew significantly. It not only complemented drawings, but also frequently became the primary method for the communication of design ideas in architecture. In particular, specialized models were made as part of the design process for major building commissions. Two such important examples were the domes of the Florence Cathedral by Filippo Brunelleschi and St Peter's in Rome by Michelangelo. Brunelleschi primarily designed in

three dimensions and used models extensively – whether as elaborate ¹⁄₁₂-scale wooden constructions for the benefit of the client, or quickly carved in wax or even turnips to explain structural ideas to the builders. Domenico Cresti di Passignano's *Michelangelo Shows Pope Paul IV the Model of the Dome of St Peter's*, 1620, perhaps best illustrates the importance of the model within this period. This painting depicts the architect using a large wooden model of St Peter's basilica to explain and sell his ideas to his papal client. The significance of the model as a method of communication is evident in the way the model is represented in the painting. The high quality of workmanship apparent shows how the proposed building could exist at full size. The model is the focal point of a conversation between the architect and the client, who will evaluate the design from it. This painting also signals a change in the function of the model during this period, from a vehicle for exploration to a descriptive tool used to explain a design.

Inspired by Antonio da Sangallo's huge wooden model of St Peter's in Rome – started in 1539 and, at ¹⁄₂₄ of the full size, taking several years to construct – Sir Christopher Wren commissioned the Great Model for his design for St

Opposite, left
Domenico Cresti di Passignano's painting *Michelangelo Shows Pope Paul IV the Model of the Dome of St Peter's*, 1620.

Opposite, right
Wooden model from around 1717 of the Church of Saint Mary-le-Strand, London, designed by James Gibbs. This type of model is typical of those made in the period between the sixteenth and nineteenth centuries.

Right
Sir Christopher Wren, model for dome of Royal Naval Hospital, Greenwich, London. This model, from circa 1699, is one of the earliest examples of a sectional study in British architecture.

Paul's in London. Built between 1673 and 1674 by a team of craftsmen, the Great Model was accurately made at a scale of 1 inch to 18 inches, enabling the client and prospective builders to walk inside its 18-foot-high interior. However, it is evident that Wren considered the model to be for the benefit of the client and builders rather than for his own design-development purposes as he wrote: 'a good and careful large model [should be constructed for] the encouragement and satisfaction of the benefactors who comprehend not designs and drafts on paper'.[6]

Prior to the eighteenth century, architectural models were primarily produced as either descriptive or evaluative devices, or as full-scale prefabrications used to predict structural behaviour. However, during the mid-eighteenth century, and coincident with the newly founded technical colleges, the use of physical models for teaching purposes became more widespread. Such models represented the more complex structural and constructional conditions that this period ushered in, and they were used in the education of technical students and building tradesmen. A major resurgence in the use of the model as a design tool in architecture can be traced to the start of the twentieth century – for example in the work of Walter Gropius, who, in founding the Bauhaus in 1919, was keen to resist the prevailing preoccupation with paper designs in favour of physical models to explore and test ideas quickly.

From this point on, the scale model re-established itself as a vital design tool for architecture. The design-development model was to have an important role in the conception and refinement of countless built and unbuilt projects of the Modernist era. For example, there was Gerrit Rietveld's attempt to give architectural form to the ideas about space that he had previously explored in his furniture designs. From the sequences of models for his design of the Schröder House, it is clear that Rietveld's starting point was a block form, whose coloured surfaces combined with receding and projecting parts to break up the massiveness of its volume. A similar intent is discernible in Vladimir Tatlin's search for a monument to represent a new social order in Soviet Russia, described in the great model of his leaning, twin helicoidal tower, the Monument to the Third International. The actual building of the model took just less than eight months and was undertaken without preliminary sketches, enabling Tatlin to explore design possibilities as the model was constructed. The progression of architecture throughout the twentieth century bore witness to the increasingly common use of models as explorative tools in an architect's design process.

Opposite left
Delmaet and Durandelle's plaster model from 1864 of the capital for the columns on the façade of the Opéra Garnier, Paris, designed by Charles Garnier.

Opposite right
Vladimir Tatlin's assistants building the first model of his Monument to the Third International from wood connected by metal plates, 1920. This was one of several models made for this project that tested various iterations of the design. A simplified version was paraded through the streets of St Petersburg (then known as Leningrad) in 1926.

Right
Sir Edwin Lutyens' proposed design for the Metropolitan Cathedral of Christ the King, Liverpool, 1929–58. The model was constructed from yellow pine and cork, and made by John B. Thorp in 1933.

Below
Model for the Concrete City of the Future, designed by F. R. S. Yorke and Marcel Breuer, 1936.

Sir Denys Lasdun and Philip Wood looking at the model of the library extension for the School of Oriental and African Studies, London, 1972.

It was this direct handling of materials and space through the use of models that heralded the key early twentieth-century architects as creative designers who visualized and articulated their concepts in a provocative and unconventional manner. A study of their formative experience and design processes reveals an explorative nature, which, being founded upon an understanding of spatial possibilities, transcends a singular reliance upon drawing. Beyond this point in history, the physical model was established as a powerful method of communication in the description, exploration and evaluation of architecture. The model has been an important method of communication in the understanding of architecture for over 500 years. Whilst the increase and developments of new technology have enabled Computer-aided Drafting (CAD) to become a powerful design tool in architecture, the use of physical models remains a key aspect of education within the discipline and for many practices around the world.

Below

Wooden concept model showing the geometrical solution for the pre-cast concrete shells, Sydney Opera House, Sydney. Designed by Jørn Utzon, this landmark project initiated an era of complex geometry in modern architecture.

Bottom

1:20 model of the design for the new additions to the Reichstag, Berlin, by Foster + Partners. Made by Atelier 36, using MDF (medium-density fibreboard), acrylics, steel and brass, this model of the plenary chamber, dome and light cone was hoisted onto the roof of the Reichstag to test the lighting in 1996.

Modelmaking now

Why use physical models to describe and explore the qualities of architecture? The most obvious answer lies, of course, in the actual tangibility of such models. Physical models enable designs to be explored and communicated in both a more experimental and more rigorous manner than other media, as various components of the project may not appear to make much sense until visualised in three dimensions. Akiko Busch suggests that part of our attraction to models lies in the fact that 'the world in miniature grants us a sense of authority; it is more easily manoeuvred and manipulated, more easily observed and understood. Moreover, when we fabricate, touch, or simply observe the miniature, we have entered a private affair; the sense of closeness, of intimacy is implicit.'[7]

The integration of digital technology with traditional modelmaking techniques has resulted in significant and exciting shifts in the manner in which we engage with the design process of architecture. The proliferation of computers and advanced modelling software has enabled architects and students alike to conceive designs that would be very difficult to develop using more traditional methods – yet despite this, the physical model appears to be experiencing something of a renaissance. This return to 'analogue' models seems to confirm that, as Peter Cook suggests, 'as we become cleverer at predicting colour, weight, performance or materiality, we are often in danger of slithering past the question of just what the composition of space may be … [since] … the tactile and visual nature of stuff may get you further into the understanding and composition of architecture'.[8] Furthermore, the application of CAD technologies as part of the production of physical models is increasingly widespread through processes such as CAD/CAM, CNC (Computerized Numerical Control) milling and rapid prototyping. The translation of computer-generated data to physical artefact is reversed with equipment such as a digitizer, which is used to trace contours of physical objects directly into the computer.

Computational modelmaking offers a different set of techniques and tools for the designer compared to traditional methods, thereby increasing the development of design innovation and the production of architectural knowledge. The tactile qualities of constructing and handling a physical model afford the maker contact with the real world and so any overlap between different techniques and media, both digital and physical, can only enrich the discourse within the discipline even further. Indeed, as Karen Moon writes, 'Even as architecture moves beyond the realm of the material, the physical model – contrary to expectation – may not lose its purpose. Models produced at the push of a button cannot offer the individuality and range of expression requisite for the task, nor can the imagination of architects be satisfied in this manner.'[9] The

potential for computer technologies to run parallel to, or combine with, hand techniques as part of the design process is an exciting evolution that suggests the days of physical models are far from over. From the current situation, the future of architectural design looks to retain physical models as essential tools, which enable design development and communication at the very core of its practice and education.

Top
CAD model of a design proposal in context.

Above
This physical model was made using CAD/CAM processes.

About this book

Already, we have touched upon a number of different ways in which models are used in architecture. Models are important objects for the communication of architectural designs, but perhaps of more significance is their application as adaptable tools as part of the design process of architecture. In this sense physical models are dynamic, as their role can change depending on the stage of the design process, who is using them and why. The purpose of this book is to provide an essential basis for readers of all levels, from students to professionals and any other aspiring modelmakers. The content of the book aims to give inspiring examples and practical explanations for those approaching the subject for the very first time, whilst offering an important source of information for those with more experience to return to again and again.

In format, the book will introduce key principles and techniques, with useful step-by-step examples, case studies and handy tips in each section. The making of architectural models has become both a profession in its own right and an independent art form replete with its own materials, tools and methods. Architecture students are typically required to learn and develop the techniques and principles of modelmaking on their own, and therefore this book presents background knowledge alongside different types of models and their appropriate application during the design process. Furthermore, the book will introduce a variety of materials and discuss their properties in addition to advice on the tools and machines with which to work with them. It is therefore intended that whilst this book may not be an exhaustive account of every potential variation of model, it is expansive and comprehensive enough to enable readers to benefit from the diverse opportunities of modelling and to transform their design ideas creatively and instinctively.

This book therefore is divided into three main parts concerning various media, types and application of models. In the first section we will introduce a variety of modelmaking materials and media, and discuss their properties in addition to advice on the tools and machines with which to work with them. The second part explains the different types of models that designers make as they develop and communicate their ideas during the design process. The third part of the book examines the ways in which models are used or applied by architects and students. This is a key aspect of this publication since although each section may be referred to in isolation, the accumulation of knowledge regarding models throughout the book is reflected in its structure. By first describing what models may be made from, then elaborating on why different types of models are produced, the final part of the book explains how models are used and by whom to broaden and deepen the reader's understanding of their role in the discipline. To best illustrate these themes, each

The combination of lighting and photography with a physical model can achieve provocative and atmospheric results.

part of the book contains an extensive range of examples from leading practitioners, both established and cutting-edge, alongside innovative work carried out by students of architecture and urban design. In summary, it is intended that, by reading and subsequently referring to these three sections, the reader will develop a clear understanding of the wide spectrum of possibilities in architectural modelmaking to inform their own practice and knowledge.

1 T. Porter, *The Architect's Eye* (London: E & FN Spon, 1997), p.3.
2 S. Allen and D. Agrest, *Practice: Architecture, Technique and Representation* (Amsterdam: G+B Arts International, 2000), p.32.
3 R. Janke, *Architectural Models* (London: Thames & Hudson, 1968), p.15.
4 C. Mills, *Designing with Models* (New York: John Wiley & Sons, Inc., 2005), p.ix.
5 R. Janke, *op cit*, p.8.
6 As quoted by Tom Porter and John Neale in *Architectural Supermodels* (Oxford: Architectural Press, 2000), p.8.
7 A. Busch, *The Art of the Architectural Model* (New York: Design Press, 1991), p.11.
8 P. Cook, 'View' in *The Architectural Review*, 1333, March 2008, p.38.
9 K. Moon, *Modeling Messages: the Architect and the Model* (New York: Monacelli Press, 2005), p.211.

Getting started

There is a set of basic tools that a modeller needs in order to make models out of paper and cardboard; these are a cutting mat, a metal ruler, knives and scissors.

A cutting mat provides a surface on which to cut materials. This is usually made of rubber as it reduces the chances of slipping when using knives on it, but a piece of hardboard or other similar material can be used as a suitable alternative. Rubber cutting mats have several advantages. They typically have a grid printed onto them that provides a guide for lines to be cut quickly, they accept cuts easily yet are very hardwearing, and they also protect the surface upon which the model is being made.

A metal ruler is essential as it not only guides the knife in a straight line when cutting but will also prevent the knife from slicing into the rule – or the modeller's fingers! By contrast, a plastic ruler will not provide this safety and in no circumstances should a scale ruler be used to cut against, because the knife will damage its edge.

Knives are important for cutting materials accurately and should always be sharp to ensure a clean cut. Care should be taken when cutting materials, as rough edges will be obvious when pieces are joined together. Perhaps the most useful type of knife is a scalpel blade, as it is very sharp and can be employed to make fine cuts and small apertures such as windows. As a result, it needs to be used with great care since if too much pressure is applied the blade may snap. The utility or craft knife is another common tool, since it can be used to cut a variety of materials in their original form. Its advantages are that it is simple and inexpensive, and that many different versions are available for different cutting tasks. When selecting a knife, it is important to make sure that it fits comfortably in the hand and that the blade is rigidly attached so that it can be used effectively and safely. Please remember that any cutting with knives should be done carefully in a well-lit environment.

Scissors have a fairly limited application when making models and are only really used with paper and thin card, but are quick and easy to use. For cutting materials of greater thickness and density, they are usually handled in a workshop environment using tools such as a bench saw, table saw or jigsaw.

Clearly, the more elaborate the model under construction, the greater the need for additional tools with which to work the materials. Therefore, it is worth mentioning at this stage some useful secondary pieces of kit that will aid modelling on a regular basis:

Tweezers – a pair of these will increase the precision of the modelmaker and enable very small components to be easily manipulated. For models with parts that are only a few millimetres in size, this is an essential tool as handling with fingers may be very difficult.

Sandpaper – this is available in different grades depending on the material and the level of coarseness required. It is used to smooth surfaces and to tidy cut edges and apertures. For more even results the sandpaper may be wrapped around a block of wood, facilitating easier movement and control.

File – similar to sandpaper, this can be used with a range of materials to finish edges and corners. Different files are available for different materials, such as metal and wood. A file typically allows for more specific and intensive work to be carried out, whereas sandpaper provides a general finish.

All the above equipment is highly useful for cutting and making components for models, but in nearly all cases we need a method of fixing elements in place and connecting them together. The most frequently used medium for this is glue – though pins, masking tape, sticky tape and other options are available depending on both the speed of the modelmaking process and the degree of permanence required of the results. There are a variety of glues available for different tasks and materials, so remember to check that each is suitable for the purpose you intend to use it for as some glues may discolour or dissolve certain materials. The main types of adhesives and their properties are outlined below:

All-purpose glue is usually a transparent, solvent-based adhesive and can be used with a wide range of materials including paper, cardboard, wood, plastics and metal. Carefully test a small area first as it may react with some materials – such as polystyrene, which it melts!

Polyvinyl acetate (PVA) glue is a white, general-purpose adhesive and will be appropriate for working with a variety of materials such as paper, cardboard and wood. It is more effective if applied to both the components to be connected, which are then pressed together. This type of glue only works with porous materials and is not appropriate for metals and plastics.

Adhesive sprays are aerosol glues that are useful for fixing lightweight card and paper. Their key benefit is that they allow components to be stuck together and then repositioned if required.

Specialist glues are available for particular materials, such as balsa, cement and plastic. Always follow the instructions carefully when using glues such as these, as they may need to be applied in a different manner to more general-purpose adhesives.

Superglue is a transparent, very fast-drying adhesive, which should be used carefully as it is difficult to undo connections without causing damage to model components.

Masking tape is particularly useful for holding model components together whilst a glue bonds, as it will not leave any marks or residue on the materials. Double-sided tape, meanwhile, can be used to attach elements together in a swift manner, but is generally suited to large surfaces rather than for connecting small components.

1 Wrapping sandpaper around a simple block can enable much smoother and more uniform results than using a sheet of it alone.

2 Basic modelmaking tools including a cutting mat, metal ruler, scalpel knife and glue.

3 A plane and a file. These tools are used to remove layers of material and produce a good level of finish to the surface of a model base or components. Typically used with wood, but special types of file are available for use with metals.

4 Basic set of tools found in workshops. From left to right: wooden mallet, pliers, claw hammer, coping saw and tenon saw.

5 Typical workshop environment, with benches and a variety of tools for handling and working materials. Always remember to wear the appropriate safety gear, such as goggles and face masks, and never use power tools or machinery without having undergone the relevant training or supervision.

Machines

These basic and affordable tools will serve the modelmaker time and time again, although more specific machinery and tools may be required to enable a wider variety of materials to be handled effectively. Indeed, the majority of professional architectural modelmakers have the same workshop tools as carpenters, including handsaws, files, planes, chisels and mallets, which all extend the methods by which materials – especially wood – can be worked. Most architecture schools have extensive workshop facilities in which a range of useful processes can be conducted. Remember that it is very important that any necessary training and supervision is obtained prior to working with such facilities. Whilst the range of resources may vary between different workshops, the most commonly found machines and their applications are described below.

Hot-wire cutter – this is used to precisely slice through thick polystyrene foam and leave a clean edge. It works by electrically heating a fine wire, which can then have blocks of material pushed against it in order to cut them. The wire's malleability facilitates complex shapes and curves to be cut into the foam. This machine is particularly useful for massing, urban and city models, as blocks can be cut quickly and easily.

Power saw – this is often used when a knife is no longer practical, owing to the thickness or density of a material. A table saw comes with a variety of blades, which make it ideal for cutting straight components from plastics and wood – and even for profiling the latter. The band saw is able to provide freeform and curved cuts of materials. Electric jigsaws can be used to provide similar results, but there is greater reliance on the user's ability as these are normally hand held.

Drill – this is typically used to make holes for connecting components, such as dowelling rods. Care should be taken to mark out holes properly prior to drilling, and a drill should preferably be used within a drill stand to ensure precision of depth and the correct angle for the apertures created.

Milling machine – this is used to work on the surface of wood. It has a variety of different bits, which enables the machine to score features into the surface of a relief model. This machine augments the type of cuts that can be achieved with the saws mentioned earlier.

Sanding machine – there are a number of ways in which model components can be sanded, but machines make such work much more even and quicker than when it is done by hand. Disc sanders allow large surface areas to be finely sanded, whilst other types of equipment, such as belt sanders, are used to smooth curved forms.

So, once a concept and design has begun to be progressed, it is time to consider how it can be visualized in three dimensions. Although only the completed building will fully communicate the spatial qualities intended, the use of models is key to investigating the characteristics of the design and enabling decisions to be made about materials and aesthetics. This is particularly true for students of architecture, who are unlikely to be able to build their designs at full scale. In this sense, a model has even greater significance as it communicates the designer's intentions and allows others to assess the ideas embodied within it. In the next section, therefore, we will look at various different media and, with reference to stimulating examples, illustrate how they can be used to best effect for the modelmaker.

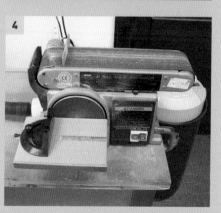

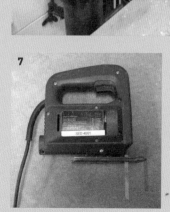

1 Hot-wire cutter: enables forms to be cut quickly and easily from thick polystyrene. Foam pieces can be guided along a straight edge, for producing block and massing models, or cut completely freeform, as illustrated.

2 Table saw: useful for cutting sheet materials for large-surface-area components and model bases.

3 Band saw: can be used to cut a variety of materials, and may be used with or without a guide in order to provide both rectilinear and freeform model elements.

4 Belt sander: used to finish curvilinear model components.

5 Disc sander: can achieve quick and even results with various model materials.

6 Rechargeable battery-powered drill: provides greater flexibility of drilling techniques but, as with all hand-held power tools, it requires ability on the part of the user in order to ensure good results.

7 Hand-held jigsaw; a versatile power tool enabling an extensive range of shapes and forms to be cut from material; however, requires some level of skill for high-quality results.

8 Drill stand: used to create accurate holes in materials using a variety of different drill-bit sizes.

MEDIA

Wooden model for Grafton Architects' Solstice Arts Centre, Navan, Ireland. The sloping ground and the geometry of the site contributed to the dynamic form of the building. The choice of material for this model reinforces this connection, as the building appears to be carved within the landscape.

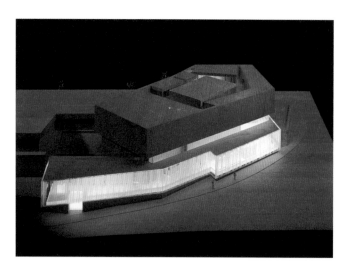

Introduction

The selection of a suitable material from which to make a model is determined by three key factors: the speed of production, the stage of the design process and the intended purpose or function of the model. However, by keeping an open-minded approach to materials, architectural ideas can be conceived and stimulated in unforeseen ways by the modelmaking experience. Experimentation with materials, especially at early stages in the design process, not only develops ideas as materials are handled and processed but also leads to different expressions of the design. This again reinforces the notion stated in the introductory part of this book that modelmaking may be equally instrumental in the generation of ideas as in their representation. With this in mind, it is worth considering the materials from which a model can be made. Each different medium has its own properties and connotations which may influence how the embodied ideas are both explored and perceived. Many people new to architectural modelmaking assume that a high degree of realism is required – similar to that demanded by the assembling of a doll's house, or a miniature train set – but in fact such detail can often be very distracting to the eye, and may not necessarily communicate the core aspects of the design. In the first instance, therefore, the architectural modeller needs to decide on the level of abstraction.

Abstraction

All models, by the very nature of their existence, display a degree of abstraction, as there would be very little point to a model if it represented reality in every aspect. Most importantly, this level of abstraction should be consistent as it would not make sense to model a building's context accurately and then insert a very loosely defined and highly abstract model of the design proposal in most situations. In essence, abstraction means taking away any unnecessary components or detail that will not aid the understanding of the design being communicated. There are no hard-and-fast rules regarding this process, and it is a skill that will develop through experience of making models, but usually the more accurate and detailed a model is the further the project is along the design-process route. This is because it would not make sense to use valuable time and resources making precise representations of initial ideas that may subsequently be very susceptible to lots of changes. As a general rule, the only constraints are what is technically possible within the time limits in which to make the model. For example, if an aperture such as a window is too small to be made it should be left out.

The greater the level of detail and communication of materials in a model, the more accurate impression

people viewing it will have of the final design intentions,
and, as a result, the majority of presentation models
made by architects and students alike are done at a stage
at which a considerable amount of the design decisions
have been consolidated. The more abstract a model is, the
more it conveys conceptual ideas and thereby allows the
designer's imagination to flow and various interpretations
of the design to be made. This is particularly useful in an
educational context, but can also assist at initial meetings
with clients and public bodies. Professional model shops
sell a wide range of parts that can be incorporated into
models – including figures, cars and trees – and these can
help negotiate the scale difference between the model
and reality. Again, care should be taken not to overly
detail a model as such information can distract from the
qualities of the design, and many designers often make
their own versions of these elements in order to give
themselves more creative freedom. The scale of the model
typically provides clues as to the appropriate level of
abstraction, and is a useful starting point.

Size and scale

In the opening remarks of his essay for the 'Idea as
Model' exhibition catalogue, Christian Hubert states that
'size and scale are not to be confused', and yet they so
frequently are (Pommer, R. and Hubert, C., *Idea as Model*,
Institute for Architecture and Urban Studies, catalogue 3,
New York: Rizzoli, 1980, p.17). Specifically, size is directly
linked to measurement and is therefore quantitative in
nature, whereas by contrast scale is relative – i.e. it refers
to a component being relationally smaller or bigger than
another component – and, as a result, is qualitative.
However, the distinction is not quite so straightforward
as we need measurements in order to be able to set up a

physical scale for a model, and as a consequence most models are built to a recognized conventional scale. So while a model will nearly always be built at a different and much smaller scale than the real building, the different elements of the model all have the same scale-relationship to each other.

Why are architectural models typically made at such small scales? The most obvious explanations are linked to time, effort and cost. Models at a reduced scale are much faster to make and enable problems with construction sequences and materials to be anticipated. The effort in making a model allows the designer to evaluate creative ideas and develop the design accordingly. To this end, the type of model required at various stages of the design process usually outlines the amount of effort required, since a basic massing model will require less investment than a detailed sectional model of interior spaces. Clearly, the cost of even the most ostentatious model is considerably less than that of constructing a real building and it allows the design to be 'interrogated', avoiding the possibility of the building being constructed with design flaws. However, beyond these rather evident reasons there is another central factor in the role of models that is connected to the discipline itself. This relates to the design-development process, and to the fact that ideas are much easier to enhance and edit when they are smaller and simpler than the real thing. The preliminary quality of ideas manifest in physical models demands dexterity on behalf of the designer, as they are dynamic tools of both exploration and communication at all stages of development.

The selection and composition of media

The selection of the materials to be used to make a model depends on its purpose, the stage of the design process and how quickly it needs to be constructed. In order to determine which materials are appropriate, it is important, as described above, to think about the level of abstraction and the scale required. If a model is to be relatively abstract, it makes sense to make it out of one material and focus on the form and mass of the design. A single material can often be manipulated and treated in different ways, so this does not necessarily limit the appearance or level of detail. Monochromatic models are very common in architecture, made from either wood or white cardboard. However, once the initial choice of material has been made, a modelmaker then needs to consider whether additional materials will be used to produce a more representative model of the building. The combination of different surfaces, colours and elements is a time-tested method of making architectural models, but care should be exercised to communicate the most important features of the design and avoid unnecessary information.

Above
Ofis Arhitekti's model for their Curly Villages project for Graz, Austria utilizes thin, coloured foam strips combined with transparent and translucent plastics to demonstrate the curvaceous and fluid nature of their design.

Opposite
This model for Holodeck Architects' proposal for the Connected with Max project in Valencia, Spain, uses acrylic sheet to great effect, as it is either left transparent, sanded to provided translucent elements or has coloured acetate applied to provide an array of volumes in relation to different functions within the overall composition.

The arrangement or composition of its components enables the construction of a model to be an adaptable and a highly useful tool for design development, as many of the assembly processes simulate those made in reality. Combining materials in order to emphasize their contrasting qualities and make best use of them requires considerable experimentation, and modelmakers should be encouraged to explore the use of novel, 'found' materials as well as to recycle packaging and other everyday objects in the search for appropriate materials. Just as is the case when laying out a drawing, thought needs to be given to how the model will be composed. With this in mind, there are a few basic issues that should be considered before the process of making commences. Firstly, what physical scale should be chosen and how will this best relate to the level of detail required in the model? Secondly, is the proposed project to be centred within the overall model – or are there reasons as to why this would not be the most suitable position, for example: specific contextual features? Finally, what are the desired interrelationships between components in terms of colour, proportion and material that will communicate the design ideas in the most appropriate way?

The making of architectural models is often perceived as a highly skilled craft preoccupied with accuracy, and whilst this may be relevant to presentation models it does not automatically result in a good model. It is important to state here that it is possible to make creative and provocative models without laborious and time-consuming techniques, and a wide range of inspirational examples are described in the following sections of this book.

This section describes the various and most commonly used materials for modelmaking. There are other materials that may be appropriated and incorporated into the modelling process, and some of these will be discussed in Part 2: Types and Part 3: Application later in this book. For the uninitiated, it is a good idea to experiment with different materials or to use them in innovative new ways. In doing so, you will learn their properties and may make some other interesting discoveries along the way. In more conventional terms, a model is a scaled-down version of an existing situation or a future building; therefore, through the practice of modelling actual surfaces are transformed into a model through a process of abstraction. However, architectural designers usually seek to retain the specific qualities of materials and the effect they might have, because in the end a building's appearance comes primarily from the sum of the materials from which it is constructed. This raises the question of which materials the model should incorporate so that it may best represent the designer's ideas and the proposed architecture. This in turn leads us to survey the extensive range of materials available to us. Some of these materials have been used in architectural modelling for a long time, whilst others are new – and, indeed, some were conceived for entirely different purposes.

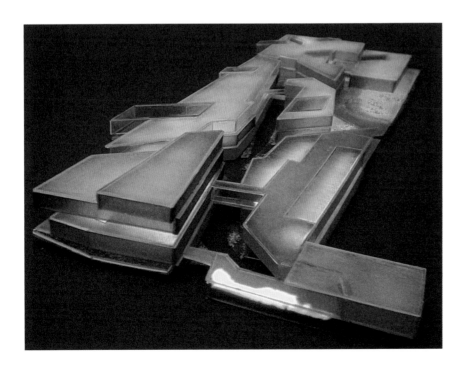

Paper and cardboard

For the designer new to modelling, the different types of paper and cardboard that are easily available and very economical make them great materials with which to start. The versatility of uses that can be achieved with these materials – through processes such as cutting, scoring and folding – means that they are ideal for a variety of different model types. Furthermore, because they are typically flexible to varying degrees (depending on their thickness) and easily manipulated, they also offer adaptability and are highly suitable for both the initial exploration of form and for detailed design work.

The flexibility of these materials is clearly illustrated in the examples on these pages, in which a number of different varieties of paper and cardboard have been used to illustrate the designers' intentions.

Perhaps a significant limitation of models made from paper and cardboard is that they are usually far more delicate and perishable than those made from more robust materials. The 'classic' white cardboard models so popular with students and architects alike is susceptible to dust and dirt, so extra care should be taken when handling them post-production and a soft, dry paintbrush applied to remove any dust. Care also needs to be taken when using a glue to fix and connect components made from paper and cardboard, as any additional residue will be difficult to remove at a later stage.

By using textured and coloured cardboard in this model, the designer has given a suggestion of how the cladding may appear. Note the relatively 'heavy' appearance of this model in relation to the example on the opposite page. Sketch modelling allows different ideas to be tried and tested quickly – in this case, exploring the external envelope of the design.

A different view of the same model. This angle reveals the 'working' nature of the model, as staircases have been made notionally and components assembled quickly in order to analyse spatial characteristics.

The use of a thin paper affords this model a translucent quality that may be indicative of the actual cladding of the building. The addition of artificial light within the model enables the viewer to perceive how the scheme may be transformed at night, and how elements of the building that may appear solid during the day become implied voids.

In this example, the designer has used similarly coloured plain and corrugated cardboard. The reduced palette of materials and colour enables the viewer to understand the 'pure' space of the model, so that it can be discussed and developed accordingly. The use of corrugated cardboard in this model raises some interesting points. Any material that is inexpensive or available for free is very practical for use in models. Primarily produced for packaging, corrugated cardboard can be 'recycled' for architectural models and has several advantages: it is easy to cut; it has a structural core, allowing it to retain both shape and rigidity; its relative thickness compared to cardboard make it useful for building up quick ground layers in site models; and its textured core affords use as yet another compositional element.

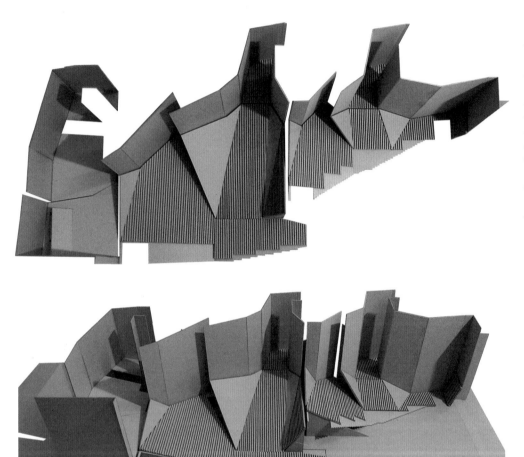

TIP CUTTING SAFELY

Always remember to use a cutting mat when working with sheets of paper and cardboard; this will prevent damage to both furniture and to you, as it will minimize slippage when cutting out components.

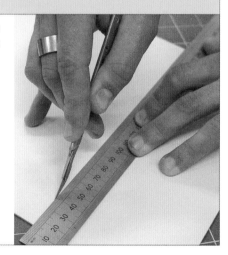

Left

The contrasting nature of different paper-based materials can offer new insight into design ideas. In this model, there has been a deliberate decision to utilize two highly contrasting colours of paper and cardboard. The darker material emphasizes the mass of the house in order that it can be studied as an object, even though it is unlikely that in the final construction materials will be as sombre and monochromatic as this. The lighter material indicates the spatial effects of the openings, such as doors and windows, and allows the designer to investigate shadow-play and internal volumes.

Below left and right

These images show how effective corrugated cardboard can be in evoking a sense of mass. The sculptural qualities of the corrugated texture are revealed, as layers of the cardboard have been attached and then the resultant solid block has been 'worked' using various machines in a workshop, including a band saw and a jigsaw. Whilst the model may not offer much in terms of detail, its communication of the relationship between mass and void combined with the aesthetic or the material chosen affords cavernous and dramatic spatial qualities.

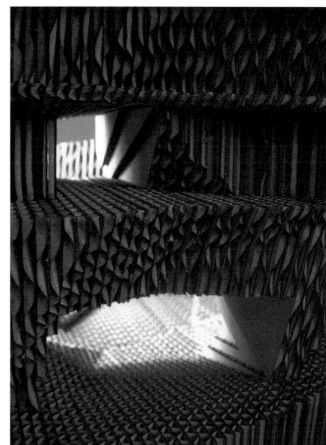

Right
This model uses cardboard upon which a printed pattern has been applied. Whilst most of the façade panels have been individually attached to one another, the resultant pattern gives the impression of a more continuous folded surface as well as evoking possible final materials.

Below left and right
A model made using folding techniques. In this case, the geometry, dimensions and fold lines are carefully worked out in two dimensions, and surface pattern applied prior to assembly as a three-dimensional object.

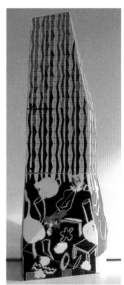

Folding in architecture has been explored by a number of different practices as they seek to investigate fluidity of space, continuous materiality or 'wrapped' buildings. Compared to most other modelling materials, paper and cardboard are available in very thin thicknesses and are therefore highly suited to folding and faceted architectural components in models.

As stated earlier, the availability, low expense and flexible properties of paper and cardboard make them ideal for use in working or explorative models, through which the designer investigates various possibilities and refines the design – whether through models of the whole scheme or smaller parts of it.

STEP BY STEP DEVELOPING A DESIGN USING PAPER MODELS

This sequence by UNStudio, made for the Mercedes-Benz Museum, Stuttgart, typifies the explorative approach, and clearly illustrates the design process and development. The design demonstrates a synthesis of programmatic and structural organizations, with the building's geometry of three overlapping circles becoming a sophisticated spatial experience.

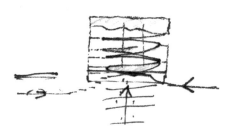

1 Concept sketch illustrating fluidity of circulation.

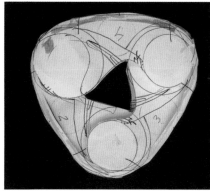

2 A quickly produced model made from paper shows the initial design idea of three circular zones connected by curvilinear areas. This model has been annotated and sketched upon as the design is developed.

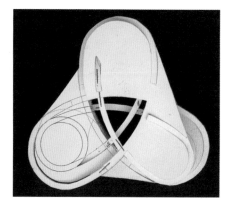

3 The development of the design is reflected in this cardboard model, which has been assembled with greater accuracy and more attention to detail so that specific elements can be explored and the geometry refined further.

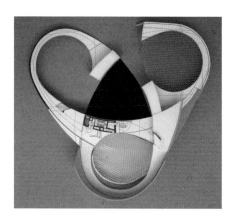

4 The introduction of additional materials in this model illustrates a consolidation of design ideas in relation to the two previous stages. At this juncture, the different zones and the access between them is more refined than before and the building's fluid spaces become evident.

Right The completed building.

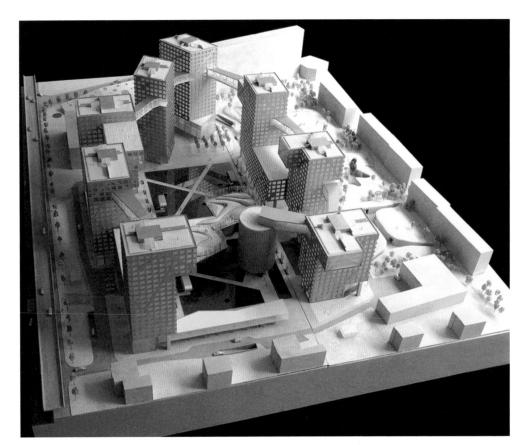

One of the significant advantages of using thin sheets of material, such as paper and cardboard, is their ability to permit the investigation of different lighting conditions. Tracing paper makes a very effective substitute for translucent glazing elements, and is easy to obtain.

The reductivist nature of white cardboard models is very popular in architecture, as they enable the viewer to examine 'pure' space. They also allow the more sculptural qualities of a proposal to be described and presented without any unnecessary colour, texture or detail to distract the eyes. These qualities were critically explored in the late 1960s and early 1970s by Peter Eisenman, whose series of residential designs were labelled 'cardboard architecture' due to their thin white

walls and model-like properties, serving to highlight the opportunity that the process of abstraction may provide as a design generator. Another useful material for this type of model is foam board, which comprises a layer of foam sandwiched between two thin sheets of card. It is available in a range of thicknesses, which makes it suitable for modelling different wall thicknesses, and, being relatively rigid, it can be used for self-supporting components or even as a base for paper and cardboard models. Foam board is most commonly known in its white variant, but coloured sheets offer further possibilities.

TIP CUTTING FOAM BOARD

Foam board is cut by applying light pressure on a knife and using multiple passes, as required, for material thickness. A sharp blade is needed, and this should be used in conjunction with a steel edge with non-slip backing. Cutting foam board will dull blades very quickly, and they must be changed often to avoid jagged edges. The blade may be angled to create mitred joints.

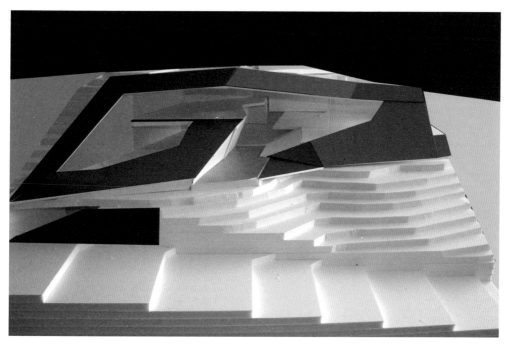

Above

Foam board is very useful for design-process models as well as presentation ones, as it enables large components to be created easily and quickly. However, the benefits of its comparative rigidity mean it can be difficult to achieve surfaces that curve in more than one dimension.

Left

In this case, the reduction of the surrounding landscape to built-up layers of foam board, upon which a basic palette of coloured card has been applied to indicate the proposed design, illustrates a typical convention in modelmaking, in which the designer wishes to draw attention to the building's imposition upon the existing context.

The ability of paper and cardboard to incorporate curvilinear geometry in two directions is significant in relation to other materials. Further experimentation with these materials can also be gained through wetting and moulding them, which not only offers different possibilities for the generation of ideas but may also have provocative implications for the actual construction process – one of the reasons Frank Gehry uses them so frequently. These materials are particularly useful at the early stages of the design process, when the form of the building is unknown and the designer may not want to waste time, effort and money testing out such ideas with more expensive and labour-intensive materials.

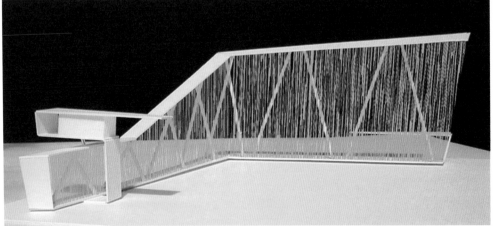

Left

The abstract nature of this architectural design is a result of the use of one material coupled with planes extending out in space. Upon first inspection, it is difficult to appreciate the scale or nature of the spaces created and how they may relate to the context.

Below

The simple addition of thin metal rods suddenly encloses parts of the model, and while the viewer may still be unclear as to the precise function, the architecture is more evident.

Opposite

These images demonstrate the ease with which paper and cardboard can be used to construct curved surfaces that can be both sleek and continuous. The subtleties of this scheme's undulating form and its relationship with the surrounding landscape is apparent, and thus, while it may not be particularly detailed, the model provides key information about the relationship between its various elements and its context. Note also the use of nails to indicate trees; whilst clearly there is no foliage evident, the nails provide basic information about other vertical elements in the immediate landscape.

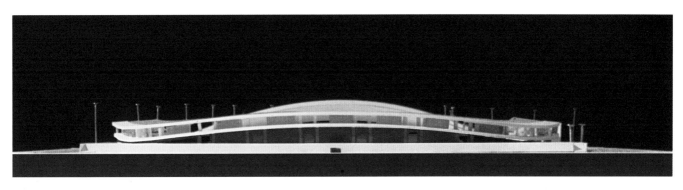

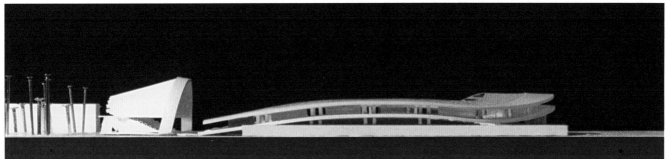

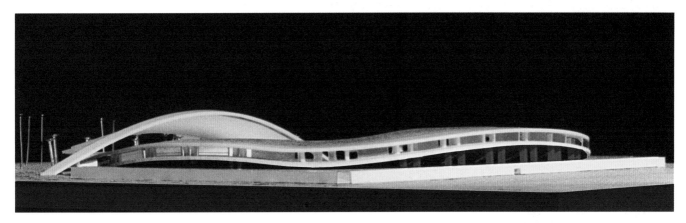

Below

For this project, folded-cardboard models were initially developed to investigate the geometry and wrapping of internal space – a process that takes careful planning to ensure the two-dimensional shape folds into the required three-dimensional form. Once the form was consolidated, textures were applied to a CAD model that bears a clear resemblance to the final building.

While paper and cardboard may be very useful for modelling, their two-dimensional qualities are usually revealed to the eye and so various painting techniques may enable the designer to give such a model a greater sense of mass or materiality. For the inexperienced modelmaker, these techniques can sometime be a little trial-and-error, and it is therefore important to always test a piece of the material separately before unleashing paintbrushes or spray-cans upon a model that may have had considerable time and effort invested in it!

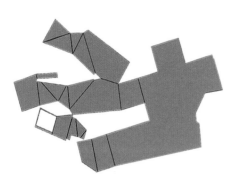

Below

The use of paint on this cardboard model gives the building a much more 'solid' feel despite its comparatively thin envelope. It also provides an indication of external materiality, whilst preserving internal spatial purity.

Left and below
Unlike previous examples in this section in which models were lit from within or underneath, in this case light is projected onto the model. The design for the Royal Ontario Museum, Toronto, by Daniel Libeskind, is based upon a concept of interrelated and fractured geometries – a common theme throughout his work – and, by projecting the lighting pattern directly onto the three-dimensional model design, the effects can be observed and decisions about the building's apertures can be made.

Left
Royal Ontario Museum, Toronto. Photograph of the completed building.

Wood

Perhaps the most established material used for making architectural models is wood, as it has prevailed for over 500 years. Unlike paper and cardboard, working with wood is typically more labour-intensive but the investment of additional time, effort and cost is often reflected in the impressive results. Since different types of wood each have their own natural aesthetic, this is usually expressed independently of the model's form and composition. The subtle variances in the grain and colour of different woods provides a rich palette with which to model. Wood can be sanded very finely and then treated in a number of ways, including varnishes and paints, but the majority of models made using this material are left uncoated since it is the natural state and appearance of the material that gives wood its aesthetic appeal.

Although a considerable spectrum of different types of wood is available, they all fit into two basic categories. The more familiar is naturally grown and dried wood that is taken and manufactured directly from trees. The other category consists of products that are manufactured from waste wood produced by timber-processing industries; these are wood-based and usually produced in sheets. Both types of wood are frequently used in modelling. An important factor when deciding which type of natural wood to use in a model is ensuring that the grain is appropriate for its scale. In most cases, it is preferable to use a wood with a smooth and relatively plain surface. Such consideration is not necessary with wood-based products as they are uniform in appearance and, as well as use for model components, are often used for making model bases or large-scale prototypes. However, great care should be taken when working with these materials (particularly wood-based ones) to ensure that that the correct protective clothing, such as goggles and a face mask, is worn, as the resultant dust can be very harmful.

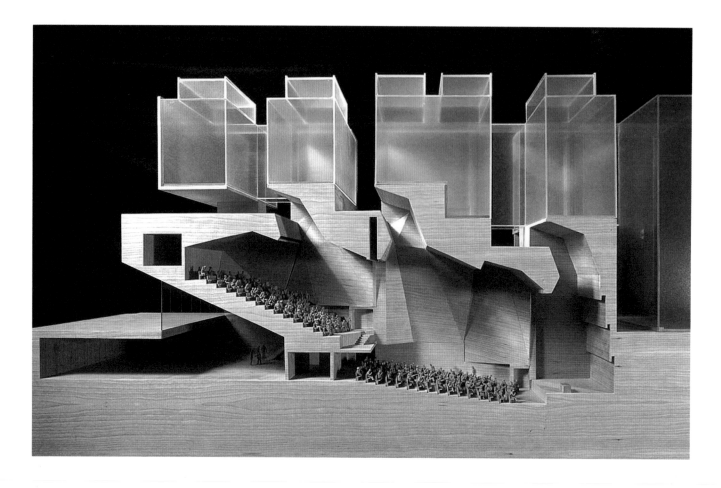

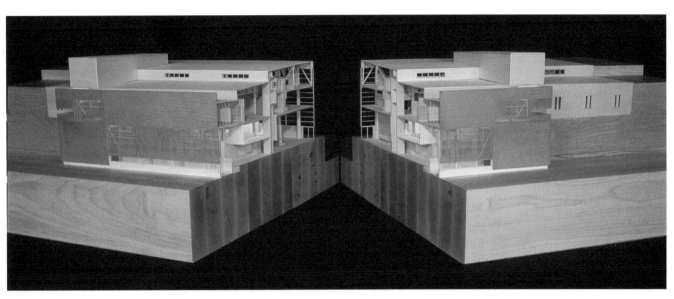

Above

In this sectional model, a different type of wood has been used to express the section and enable the viewer to look inside the spaces.

Right

This model uses a single type of wood to represent the existing building whilst the new interventions are coloured and combined with clear acrylic components, making them easily recognizable in relation to their historic context. This approach is often applied in models to enable the viewer simply and effectively to identify the new design elements within the extant conditions.

Opposite

The use of wood in this model enables the designer to emphasize the difference between the solid elements of the building, with its seemingly carved spaces, and the delicate glazed voids above.

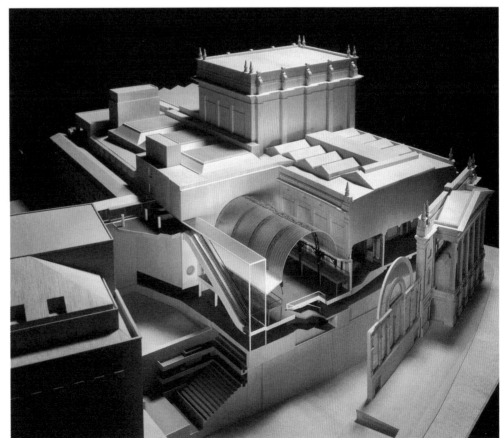

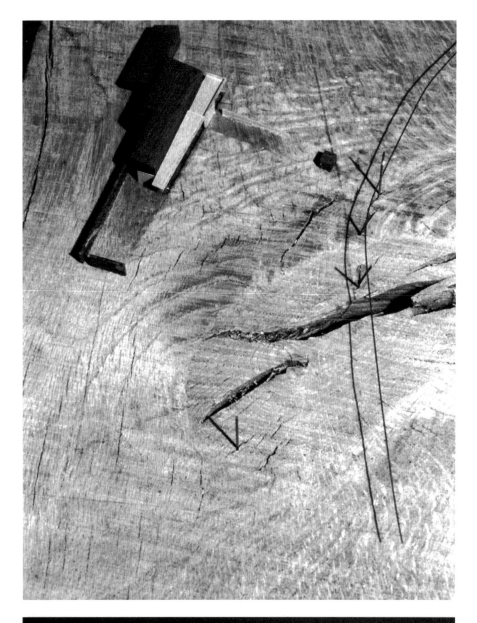

Left

For this site model, the 'raw' state of the wood used for the site is left untreated and contrasting coloured wooden blocks are used to represent the building proposal. This example is effective as the grain of the wooden landscape-base is so out of proportion with the scale of the model that we do not confuse it with actual terrain features.

Below left

This massing model, produced by stacking roughly cut pieces of softwood, enables the designer to test various iterations of the blocks – both as individual towers and as elements of the overall composition. Each variation can be photographed to record the design process, and when a desirable result is reached the pieces can be glued together to provide a more permanent model.

Opposite

This model by 6a architects not only represents the proposed design but also investigates the actual process of production, as it uses the same CAD/CAM technology (Computer-aided Drafting/Computer-aided Manufacturing) as the final architectural intervention – albeit on a smaller scale. The thin timber veneer lends itself very well to the delicate tracery within the design, facilitating the layering and relief of shapes. Note the use of a person to indicate human scale, and the contrasting but uniform manner in which the context is expressed – an approach that maximizes the impact of the design.

Left
This concept model expresses the form and arrangement of the building blocks whilst remaining highly abstract. The lack of a baseboard to 'fix' the components reinforces the playful aspects of the initial ideas, as the blocks appear movable and the possibility of numerous configurations is enticing.

Below
By using the same material for the model base and the building in this example, the designer emphasizes the integration of the design with its immediate surroundings. Note the use of horizontal grain to evoke a sense of landscape and strata in the base, whilst finer wooden components communicate the more detailed elements of the building's façade.

Right

The contrast between the solid and voids of this design is reinforced when the model is artificially lit and the some of the wooden elements appear to float. Considerable care has been taken when photographing this model to ensure no background elements detract from the effect.

Below

The sculptural quality of this design is clearly evident as a result of the 'dialogue' between the wood and sectional cut chosen. Applying the horizontal grain of the veneer to the sectional cut serves to draw the eye to the 'carved' internal spaces, and allows us to focus on the details of the interior. The effects of the glazed rooftop components can be observed within the internal spaces, with artificial lighting used (out of the camera shot) to reinforce the contrasting areas of shadow.

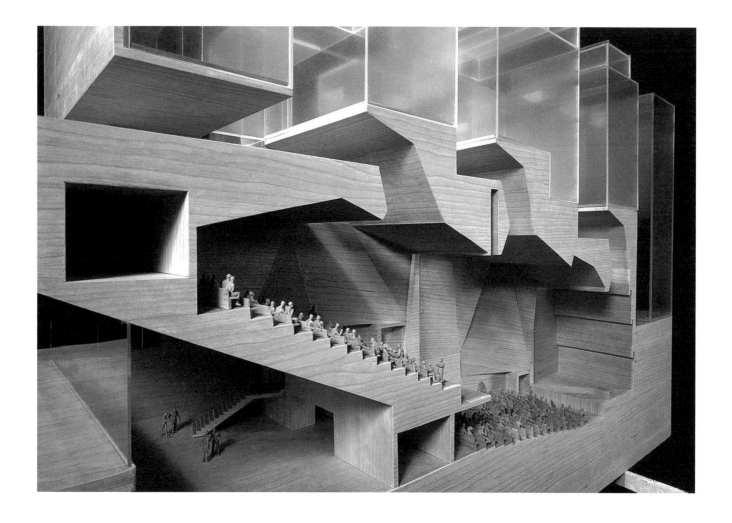

A wide variety of different woods is available and, whilst it is not necessary to list every potential type, the main ones are described here along with their properties.

Naturally grown woods

Balsa – this is a very lightweight material with a much lower density than other hardwoods, making it easy to work with. It has a coarse, open grain, and for modelling purposes it typically comes in sheets or long rods that are either square or circular in section.

Basswood – similar to balsa in terms of lightness and easiness to cut with a knife, but with a tighter grain and finer finish, this is the most commonly used modelling wood in the Unites States.

Beech – this is a strong hardwood with a fine grain, and is light brown in colour. It is available in a range of thicknesses from thick blocks to thin veneers, making it a very suitable modelling material.

Cedar – is a type of softwood, with a fresh, sweet odour and reddish colour. It is an easy material to model with, has a consistent texture and is resistant to decay.

Cherry – a close-grained hardwood that resists warping and shrinking. One of its distinguishing characteristics is that exposure to sunlight causes it to redden further.

Cork – a comparatively light material that is manufactured in a range of thicknesses, making it a flexible finish for models as it can be used with curvilinear surfaces. Its pattern and smooth finish make it particularly suitable for modelling landscape and topographical features.

Mahogany – a hardwood historically used to make furniture, this is fine grained, dark reddish-brown in colour and very durable. Veneers of this material are more commonplace in contemporary modelling than solid blocks.

Maple – a fine textured, light-yellow hardwood. It is very strong and hard, and used both as blocks and veneers in models.

Pear – this hardwood is pink to yellow in colour. Its fine surface and appearance make it a popular choice among professional modelmakers, either in block or veneer form.

Pine – a softwood with a light yellow colour and a uniform surface texture. It is easily worked with saws or even a craft knife, depending on the thickness.

Comparatively inexpensive in relation to hardwoods, it is often used for building components, such as internal partitions, in the construction industry.

Rosewood – a very hard material, close-grained and with a dark reddish-brown colour. A very dense hardwood, it can be difficult to work with – but it takes a high polish.

Walnut – this is a variable-textured hardwood with a dark brown colour. Its fine and dark appearance provides a good contrast with lighter-coloured materials, and it is easy to work with – making it a useful modelling material.

Wood-based sheets

Particle board – an inexpensive material made by gluing together wood particles under heat and pressure. It has a relatively rough surface and is available in a number of densities. With the exception of the high-density variety, it has a tendency to soak up water, making it swell and break down. It is typically used for making baseboards for models.

MDF – or medium-density fibreboard is hard and has a very smooth and uniform appearance. Its original light-brown colour can be readily changed using paints and lacquers. Considerable care should be used when cutting this material, as the resultant dust can be very harmful.

Plywood – this is a laminated sheet material produced by gluing thin layers of wood together. The type of wood used to manufacture it usually determines its grain and colour. This is an easy material to work with and is often used for baseboards and larger models.

STEP BY STEP WOODWORKING

The variety of processes for which wood may be used in modelmaking is ever-increasing due to the development of CAD/CAM techniques and the integration of these with more traditional methods. Using manual tools is comparatively straightforward but care should always be taken when using any tool and appropriate safety gear should be worn at all times. In this example an important aspect of this model series is the integrated wooden base which was specifically designed as a prototype piece of furniture in its own right as well as to support the acrylic layers of the models above.

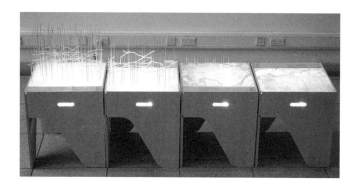

1 CNC milling bed works from a CAD fill to cut and rout the model elements from a sheet of plywood. The geometry of the design has been optimized to reduce the amount of waste material.

2 Once the model elements have been cut out they then have specially shaped slots cut into them using a 'biscuit jointer'.

3 Oval-shaped pieces known as 'biscuits' are covered in glue and then inserted into the slots. The biscuits are highly dried and compressed components of wood, typically made from beech, that expand in contact with the glue, providing a very strong bond.

4 The model elements are then clamped together whilst the joints bond. Once the clamps have been removed, the prototypes can be treated with paint, varnish or have lighting installed as shown here.

Below

This model uses a thin veneer of wood as a continuous surface, from which the building 'emerges' seamlessly out of its surroundings. This results in the geometry of the plan becoming the main focus of attention, and also reveals the subtle nature of the building's section in relation to the immediate landscape. Note the deliberately minimal palette of materials used for this model, with the effects of shadow revealing the design's composition.

Right

One of the benefits of wood-based materials is that because they are produced in sheets, model layers can be built up quickly and with minimal waste. In these examples, CAD/CAM software has been used to design the overall form, which is subsequently cut as a series of pieces that are then assembled as shown.

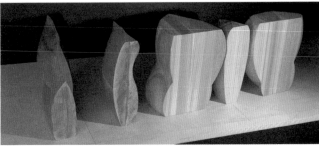

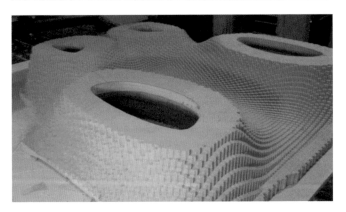

TIP SOURCING WOOD

A wide variety of good-quality wood is available at timber yards and in small hardware shops, which may provide more inexpensive alternatives than professional model shops.

Above and above centre

As well as lending itself to manipulation by traditional techniques such as carving, wood is a particularly suitable material for various CAD/CAM processes, including CNC (Computerized Numerical Control) milling.

Styrofoam and plastics

There is such a variety of different plastic products available that it is difficult to generalize about their properties and aesthetics. However, the majority of plastics are malleable, synthetic materials made of macromolecules. They share some key characteristics, as all plastics can be processed easily and with a high degree of accuracy. Their rigidity combined with lightweight properties makes them highly useful in models for which other materials would not be suitable.

One of the most common plastics used in modelling is polystyrene, as it is mass-produced and inexpensive. This makes it ideal for a range of uses. Some architects and modelmakers only work with polystyrene, particularly when developing ideas, as the material can be worked with quickly and easily. Consequently, a special type of design and presentation has emerged for constructing architectural models. Polystyrene is typically white and smooth, although blue styrofoam and pink rigid foam insulation are also common, and therefore ideal for detailed work and complex, organic forms. Its finely textured surfaces afford a smooth finish, and models produced using this material alone offer a degree of abstraction, allowing the formal qualities of the design to be fully appreciated.

A high degree of precision can be achieved when working with plastics – a feature which frequently sets them apart from other materials. It is possible to make components to an accuracy of fractions of a millimetre,

which is a great advantage for numerous different types of models and makes it highly suitable for some CAD/CAM processes that will be described later in this section. Of particular significance is the fact that plastics can be used to represent transparent components – for example, glass – by incorporating thin sheets or foils of transparent polyvinylchloride (PVC). Plastics are thus a vital material for architectural modelling. Another frequently used plastic in architectural modelmaking is acrylic glass, often known as Plexiglas. As it is a thermoplastic material it has excellent thermal ductility, making it highly flexible and thus suitable for a variety of applications. Furthermore, many of the varieties manufactured are ideal for representing glass and other transparent building elements. Their surfaces can be modified, for example satin matt effects can be achieved by grinding acrylic glass with fine-grain sandpaper, and grid structures and patterns can be produced by milling, grooving or scoring acrylic glass with a sharp blade. Whilst the number of different types of plastic may be too exhaustive to discuss in detail here, the main criteria for the application of plastic in a model should be the form and colour desired and its surface properties.

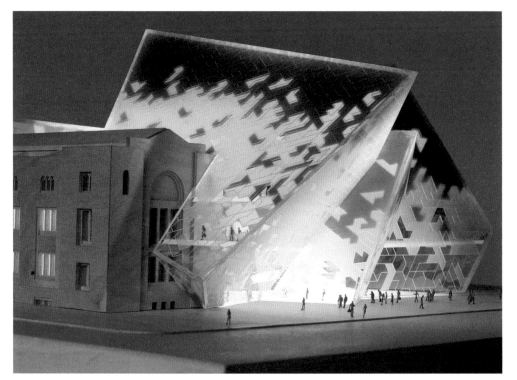

The crystalline nature of Daniel Libeskind's proposal for the extension to the Royal Ontario Museum in Toronto is reinforced in this model by the use of transparent and translucent acrylic glass. The material provides sharp lines – key to the original design concept – and evokes a striking counterpart to the existing building, here modelled in wood. The acrylic glass also enables a pattern of cladding panels to be etched onto it, assisting the viewer's understanding of scale whilst further emphasizing the contrast with the traditional architecture. (For a photograph of the completed building see page 43.)

Left
In this early design-development model for the CCTV (China Central Television headquarters) by the Office for Metropolitan Architecture (OMA), the effect of 'stacking' the various components of the building's programme, as represented by the blue styrofoam blocks, is explored in relation to the proposed external envelope, which is indicated by the continuous folded Plexiglas strip. OMA frequently use this material for their design-development work, as ideas can be investigated and revised very quickly with it; Rem Koolhaas has stated: 'often, my most important role is to undo things'.

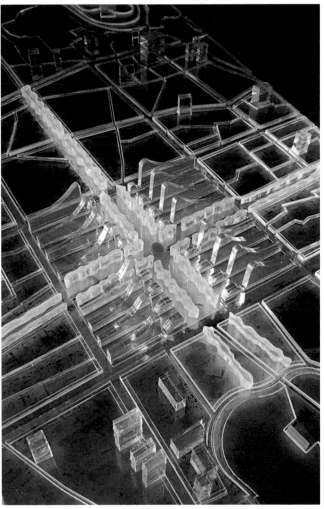

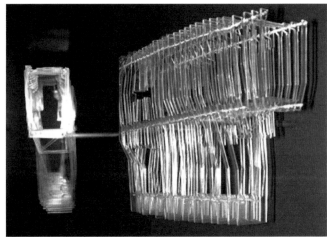

Left
For this masterplan model, 'live edge' and clear acrylic sheet (the former being a term for coloured fluorescent acrylics) was laid over the top of a colour acetate satellite image of the city to give a 'ghosting' of the urban context. The main green areas were then added, and sections of red plastic rod were used beneath to indicate the underground transport system. Finally, the main area of blue acrylic indicates the masterplan build envelopes. The model was only A4 size but photographed well, with low-angled lighting from each side which was picked up by the acrylic.

Above
The accuracy that can be gained using plastics is clearly demonstrated in this concept model, which consists of a series of similar parts carefully mounted so that, whilst it may be possible to read some spatial qualities into the model, its sculptural properties are much more prominent and we may be less certain about the intended scale by looking at the object alone.

TIP CUTTING PLASTIC

Cutting thin sheets of plastic needs to be done carefully, and it is useful to use the grid of a cutting mat for reference and score slowly several times rather than try to cut through the material in one stroke. Always remember to use a metal ruler – or a similarly hard material – as a cutting edge when cutting or scoring curves.

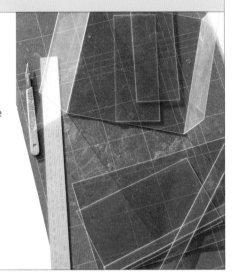

The variety of different coloured plastics available makes them ideal for incorporating into models, as and when required. In this highly playful design by Alsop Architects, the different elements are distinct and remain fairly abstract. The use of scoring on the transparent plastic façade to indicate component sizes is very useful, as it helps the viewer to understand the scale of the model without unnecessary and complex detail.

Case study Using plastics and styrofoam

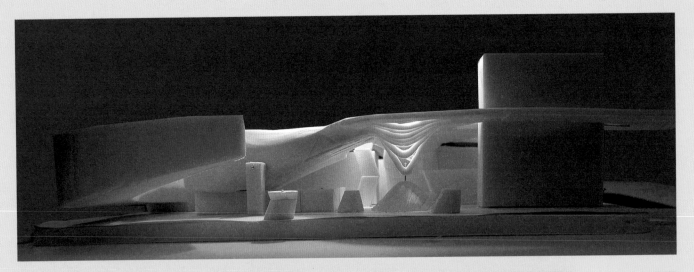

Using plastics and styrofoam to convey sculptural forms, Coop Himmelb(l)au have explored the fracturing of architecture and the amorphous nature of the contemporary city for a number of years. Key to their philosophy is the dynamism and concentration of spatial qualities that are visibly manifest in their designs. This agenda brings with it a need to communicate such aspirations, and the use of plastics and styrofoam is frequently found in their models as these materials easily embody the fragmentary and sculptural nature of their architecture. The transparent acrylic glass enables the typically complex internal spaces to be appreciated through the building's frame, and allows light to permeate them in order to add further dramatic effect.

Above and top
Design development and presentation models for the competition to design the House of Knowledge, City of Sciences, Belval, Luxembourg.

Opposite page
The presentation model (top) for BMW Welt, Munich, is shown in comparison to the final building (bottom). The practice's quest to balance experimentation and rigour typifies the dual role of models as generative and representative tools in the design process.

Right

1:1,000 presentation model made by Rogers Stirk Harbour + Partners for the Venice Architectural Biennale 2002. At this time, the practice's recently acquired laser-cutting machine enabled more intricate, stacked, clear-acrylic context buildings and bases to be made. On each stacked layer, a deep line is etched by laser inside the cut edge in order to stop solvent adhesive running across the whole face of the layers as they are glued together. The base here is made from stacked acrylic frames which form a cube with a hollow inside to take a spotlight, up-lighting the fluorescent coloured-acrylic tower scheme. The use of coloured acrylic glass in this model enables the proposal to immediately stand out from its more subtle surroundings.

Below right

Competition model for Fourth Grace, Liverpool, made at 1:500 scale. Laser-cut coloured fluorescent (also known as 'live edge') acrylics are used to allow the viewer to see through the scheme in order to appreciate the spaces and objects within. This experience is augmented by the up-lighting built into the underside of the model. Given the size of the project, plenty of human figures are used at ground level to give scale and animation to it.

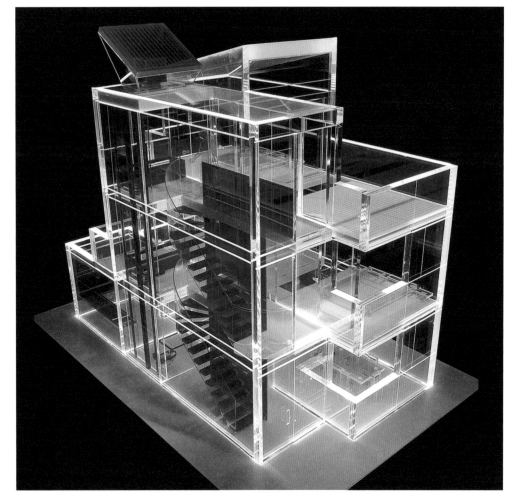

Left

The clear-acrylic wall construction enables the interior spaces of this house to be easily viewed. The coloured fluorescent ('live edge') acrylic elements within the model highlight the division between living spaces and service core, reflecting the modular construction of the actual house. By designating key components of this model with contrasting colours, the articulation of the design is heightened. For example, the vertical circulation in red appears much more dramatic than it would if the whole model had been made from colourless transparent material. Such elements help anchor the design and enable its intentions to be understood. When photographing images, a longer exposure is used to capture the integral lighting effect.

Left

This model uses thin translucent plastic to build a large sketch model of the design, in order that its internal structure and organization can be explored. This material is useful for modellers as it comes in thin sheets that allow components to be readily cut out using a craft knife, which may be much quicker than setting up workshop equipment to cut sheets of greater thicknesses.

Above

The translucent plastic used, and the method of lighting chosen, reinforces the delicate nature of this model. Instead of appearing as a potentially heavy mass, the upper section of the scheme seems to float over the more horizontal, linear and solid elements that connect it to the ground.

STEP BY STEP MAKING AN ACRYLIC MODEL

Acrylic has several advantages and therefore allows a number of design opportunities when making models. Because it is available in transparent or translucent varieties it can be placed over printed images to enable layering of further information in a model. This layering may be further enhanced by CNC cutting and etching (to which the material responds well), thereby allowing greater detail and intricacy in the modelmaking process. Different effects may also be achieved by careful sanding with a fine grade of sandpaper, or with colour applied to the newly textured surface using ink or paint as appropriate.

1 A block of transparent acrylic is cut to the dimensions required.

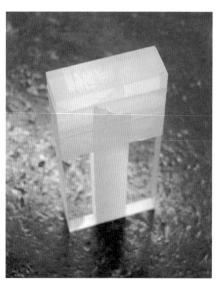

2 Masking tape is carefully applied to the desired areas, and the block is then sandblasted.

3 When the tape is removed, it has both transparent and translucent properties and it can then be incorporated into a model. A similar effect can be achieved using sandpaper or sanding machinery.

Left This model uses several layers of acrylic sheet that has been etched and cut to communicate different strata of information. The integration of contextual data underneath, along with the use of live edge acrylic to emphasize key components of the model, serves to further this.

Left

In this example, thin layers of foam have been built up quickly by cutting various pieces and pinning them together in order to test a combination of the forms. The flexible nature of this material allows for revision and reassembly of the stacks prior to the consolidation of design ideas, which will lead to modelling them in a more permanent manner – for example, using wood (see page 46).

Below left

The styrofoam of this model enables the sculptural qualities of the design to be explored. Rather than an overtly architectural proposition, the scalpel marks and hewn surface of the material give the model the appearance of an artist's maquette or mould. The inexpensive nature of this material makes it very suitable for such initial formal experimentation between mass and void – and because it can be worked very quickly, using basic tools, it allows the designer's imagination to explore many ideas. A much smoother version of the model was later produced from a mould using resin (see page 66).

TIP CUTTING PERSPEX

Perspex blocks can offer greater resistance than wood to saws when cut, and it is often helpful to use a piece of scrap wood as a 'pushstick' to guide the material through the saw and reduce the risk of injury.

This page and opposite
For this model of the Seattle Public Library by OMA, different plastics are used to communicate different aspects of the design. In the image to the left, the three-dimensional organization of the building's programme is explicit, and the fluid circulation between the 'floating' main elements is clearly indicated. By contrast, in the image below, the geometry of the transparent outer envelope wraps around the building programme and gives a sense of the external appearance. The photograph opposite shows the completed building.

TIP COLOURING CLEAR ACRYLIC

Sourcing the exact colour for translucent elements is not always possible, so a fine dusting of spray paint over clear acrylic sheets can be a good substitute. This will also provide you with a component that has one glossy side and one that is relatively matt in appearance, which may contribute to the overall composition in a model.

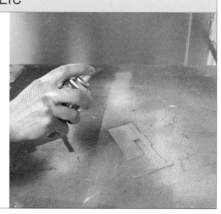

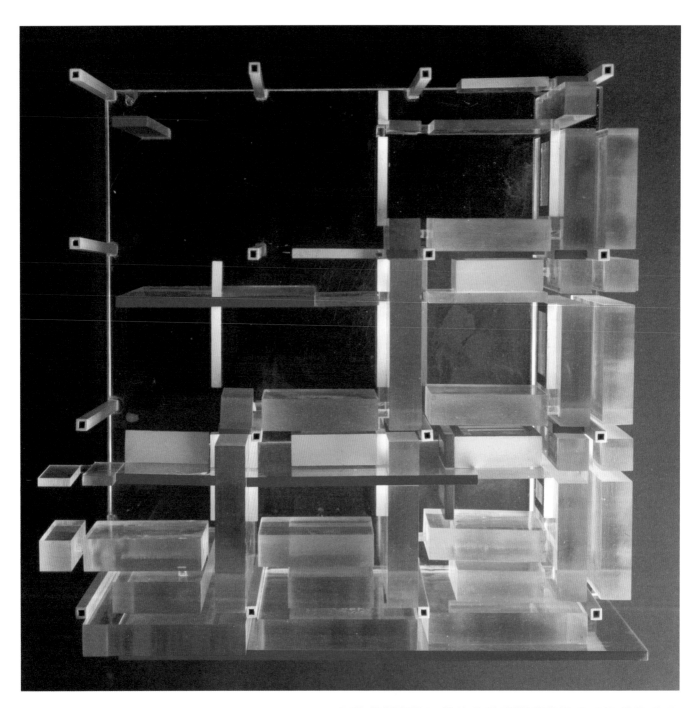

The use of transparent acrylic glass with coloured ends transforms the spaces of this design into a three-dimensional Mondrian-style arrangement. Whilst this image shows the model in plan view, its high level of abstraction provides very few clues as to the actual architectural characteristics but instead offers a geometrical analysis typical of the earlier work of its creator, Peter Eisenman.

TIP SPRAY-PAINTING ACRYLIC

Painting clear acrylic model components can be difficult, and the shiny and smooth finish of the material has a tendency to reveal brush strokes. Therefore, it is recommended that the desired areas be carefully demarcated with masking tape and then coloured using spray paints. Remember to keep the nozzle of the spray paint at a suitable distance from the material, and to build up layers of colour gradually and in an even fashion.

Resin, clay and cast materials

Cast materials are typically very malleable, and this adaptability is retained to various degrees depending on the particular material and how it is subsequently treated. This category includes materials such as clay, plasticine and plaster – the latter also commonly known as gypsum. Modellers only usually work with one of these materials in a model at a time, since their different properties do not make them easy to combine. The 'hands on' nature of this type of materials means that architectural models made from them tend to focus on exploration of form and relationships between mass and void rather than incorporate detailed design information. Despite this, it should be pointed out that it is possible to cast plaster with considerable accuracy if required, in order to produce complex and fluid three-dimensional forms with a smooth finish. In the construction process gypsum, also known by its chemical name of calcium sulphate, is used at full-scale – typically, though not exclusively, for internal finishes. Working with this

material for architectural models involves producing a mould for the liquid plaster to be poured into. Although this can sometimes be a time-consuming activity, the benefit is that once the mould has been produced it can be used over and over again as necessary. This is particularly useful if a designer needs to produce a lot of repetitive components for a model. Modelling plaster is a white powder mixed with water for use in liquid form as described above, or may be used in thicker consistencies with craft tools. Whilst plaster in its liquid state can sometimes be messy to work with, it quickly sets and hardens. When it has transformed into its solid state, it can be worked on further with a craft knife or sandpaper for more detailed work on surfaces – and it may be painted or varnished to extend the possibilities of its appearance.

The roughcast texture of this initial design model enables the effects of light and shadow to be studied across its various projections and indentations. When placed upon a material that is smooth and highly contrasting in tone, the conceptual materiality of the scheme is further highlighted.

Below left and right

A resin model produced using a mould. Compare the smooth geometry of this model to the one produced earlier in the design process using styrofoam (see page 61). With this model the impact of ALA Architects' proposal for the Warsaw Museum of Modern Art can be investigated more rigorously by the designer and presented to others. This model was then coated with metallic paint to further demonstrate the sculptural qualities of its fluid form. The lack of surface detail is a deliberate decision to emphasize its formal purity and mass.

The use of clay as a modelling material is long established across a range of creative disciplines. For the purpose of architectural models, clay has often been used to sculpt three-dimensional ideas quickly and to investigate form – especially organic forms, which may be difficult to model in other materials. Perhaps more than with any other modelling material, the tactility experienced when using clay promotes a direct engagement with it that encourages further ideas to be explored. Such sketch models can be developed to test design ideas simply and effectively. The most important factor when working with clay is to keep it suitably moist by the regular addition of water to ensure it does not crack or dry out. Once the model is ready it may be fired in a kiln to consolidate its form, the practice more familiarly known as 'ceramics' or 'pottery'. A similar process is used on a much greater scale to manufacture bricks and roof tiles for the construction industry.

Further to these materials, plasticine and air-drying modelling clay are also used in architectural models. Plasticine (also known as plastic modelling clay), unlike the previous materials, remains formable and can be continuously reworked as it does not dry out or harden. This makes it ideal for design-development models, in which ideas can be repeatedly tested, documented and then revised. It is available in a variety of different colours that can also be mixed together to provide further and subtler variations. Air-drying modelling clay, as the name suggests, has characteristics very similar to clay but does not need to be fired in order for it to harden.

STEP BY STEP CASTING A PLASTER MODEL

Plaster is an inexpensive and interesting material to experiment with as it affords the modelmaker the opportunity to produce model bases and components that incorporate topography and textures during the casting process. It may then be worked on further using knives, sandpaper or paints depending on the desired finish. This method has an almost limitless range of possibilities but it is useful to have an idea of the intended application in terms of scale and size of the surface and component to be cast.

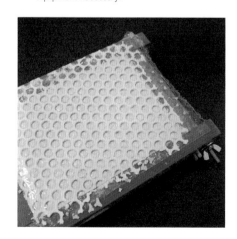

1 A wooden frame is made and lined with the desired material – in this case, bubble-wrap – which will provide the textured finish. Plaster, water and a spoon for mixing are the only basic equipment necessary.

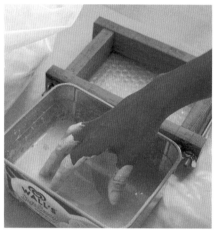

2 Following the manufacturer's instructions, the powdered plaster is added slowly to the water and mixed to ensure a correct and even consistency.

3 The liquid plaster is left to set. Times may vary depending on the consistency and type of plaster, so always refer to instructions provided.

4 When dried, the frame can be inverted and carefully removed.

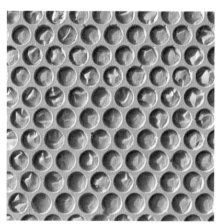

5 The resultant textured surface can be used as required – for example, as a model component or the mock-up of a detail.

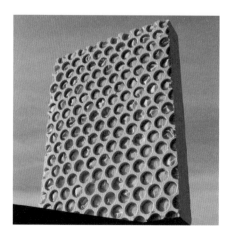

6 In this example, the mock-up cladding panel can be photographed in natural daylight to evaluate its appearance.

Below

Cast materials, such as plaster, allow relationships between solid and void to be investigated in a different manner to sheet-based materials, such as paper and cardboard. The cohesiveness of their continuous surfaces directs our attention to the sculptural qualities of the architecture. Further detail can be carefully carved into the surfaces and mass of these models, dependent on the level of information required.

Opposite page

The beauty of these smooth, fluid forms is represented in this model using very carefully produced and finished plaster components. The secret to making curvilinear forms such as these from plaster is to spend plenty of time carefully constructing precise moulds, as any flaws will immediately be apparent after the casting process. When lit appropriately, the material can give the appearance of sculpted, in situ, cast concrete – as is particularly evident in the top image here. This modelling process also replicates a similar sequence that occurs in the construction of in situ concrete itself, in which accurate moulds (or 'formwork') need to be made prior to pouring. Even the most carefully constructed mould can occasionally leave rough edges or minor blemishes on plaster. To treat these, use a fine grain of sandpaper and smooth off with slow, even movements in one direction.

TIP WORKING WITH CLAY

Using air-drying modelling clay can save lots of mess and, as an added advantage, it can be worked using basic equipment such as a cutting mat and knives.

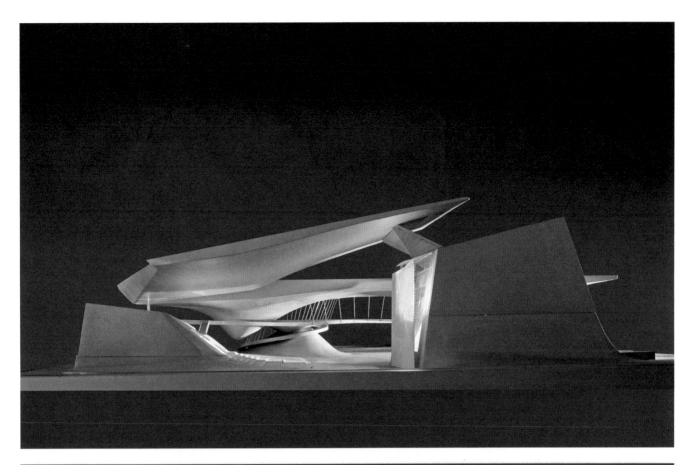

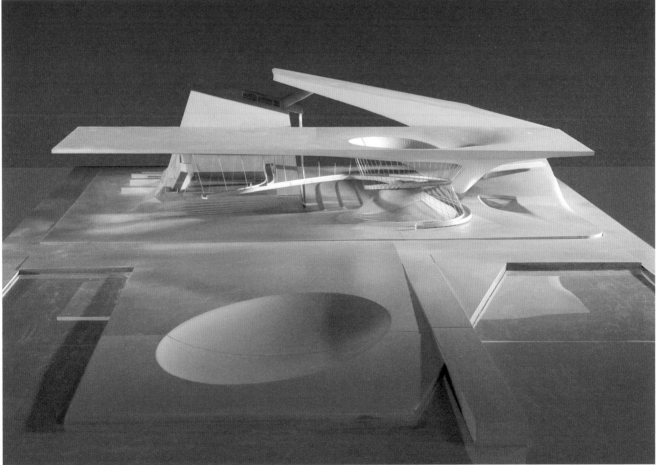

Right and far right

The transparent properties of resin facilitates the use of coloured pigments to produce component variants, as seen in these images. The resultant translucency permits the relationships between the interlocking elements of this design to be clearly communicated (left), and further described using different lighting conditions (right). This is a key stage of the design process that Neutelings Riedijk, the architects for this project, employ, as they are interested in the way in which the various functions within a building stack together and relate to voids and public space.

Centre right

The different components of the model shown above, all of which were cast separately. Owing to the interdependent nature of the pieces in this model, it is essential that great care be taken when constructing each mould, in order to ensure a close fit when they are removed and fitted together.

Below right

This close-up image of a single component of the same model enables the extent of the detail and the precision of the casting process to be appreciated. Note how some surfaces in this piece have been carefully sanded to enhance the material's qualities and to create variation within a single element of the model.

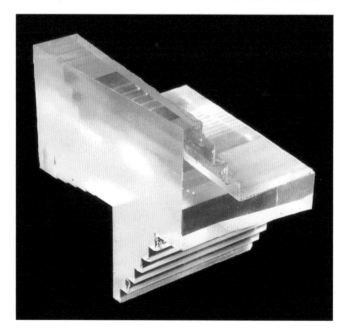

Opposite top

This roughcast plaster model showcases how gypsum can be used to project a sense of materiality. The mass and variety of forms within the design are immediately apparent, and the addition of controlled lighting allows the formal qualities to be explored even further.

Opposite bottom

The same model viewed from the other side demonstrates the rich composition of volumes and their interplay with light and shadow – as elements which were recessed in the above image suddenly project outwards, and vice versa.

STEP BY STEP USING PLASTICINE TO DEVELOP FORM

In the modelling sequence below, it is possible to see the different stages of the design process for the development of the form of ARCAM, Amsterdam, by René van Zuuk. The sculptural shape of the design is evolved through a process of contouring a piece of plasticine from a basic trapezoidal block to a much more curvaceous and sleek form. This design development illustrates one of the fundamental approaches in architectural design as it is the result of a quest to balance contextual limitations while pursuing formal inquiry. For more images of this project see pages 118 and 124.

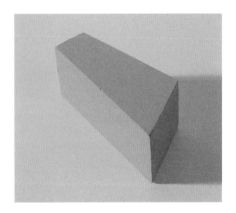

1 A basic block is cut from the plasticine.

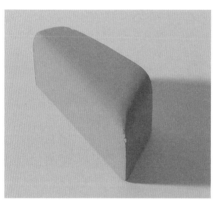

2 The edges of the block have been curved, and the form has a smoother overall appearance.

3 The form is streamlined even further, and it is possible to see the emergence of a large aperture or feature as the form is also chamfered at the front.

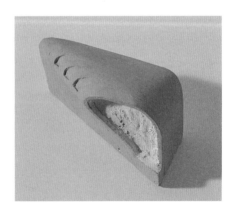

4 A refined version of the building's form, in which the front aperture is more developed and the model has more detail across its surface in order to describe its appearance.

Right A photograph of the completed building.

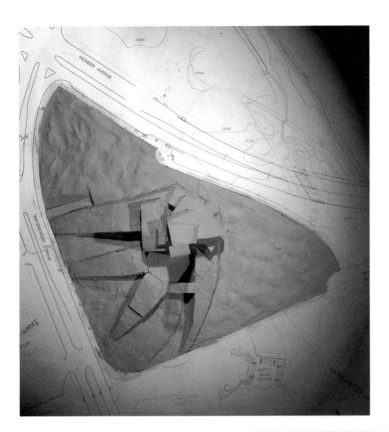

Left and below
The use of plasticine in these models allows the designer, Antoine Predock, to quickly develop ideas and revise the design as desired. This material is especially useful for modelling topography as it can easily incorporate complex site geometry in three dimensions, often by using fingers to make impressions or build up layers. In addition, plasticine is useful for projects in which the landscape is an integral part of the design proposition as it may be readily adapted and, if one type of the material is used, appear harmonious with the building.

STEP BY STEP CASTING A CONCRETE MODEL

Of course, one of the main materials used in the building industry is concrete. This is usually pre-cast or cast in situ, depending on the design, cost and sequence of construction adopted. It can be a useful exercise to test design components by working with the actual materials directly, where available.

1 Softwood blocks are cut at suitable lengths, to be assembled as shuttering for the concrete to be cast into.

2 A simple wooden frame is made. The frame can be lined with heavily grained wood, so that the pattern of the timber shuttering is revealed in the final surface of the building.

3 Additional material can also be used to line the frame and allow experimentation of surface texture – in this case, folded polythene sheeting.

4 The concrete is then mixed and the frames are filled. Time should be taken to systematically and gently press the mixture down into the frame, in order to ensure even casting and to avoid air bubbles.

5 The concrete is allowed to set according to manufacturer's instructions.

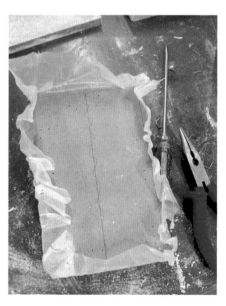

6 Having removed the frames and shuttering, the concrete component is left – sandwiching the textured material within.

7 The two layers of concrete are separated, revealing the plastic lined element.

8 The plastic sheeting is carefully pulled away from the concrete surface.

9 The resulting concrete panel can be used to examine the textural possibilities of full-size building components, or this technique could be used to form elements and bases of models across a range of scales.

Steel and other metalwork

For the purpose of architectural models, metal is typically used in sheet form to represent cladding or other building finishes, although metal rods, sections and mesh may be incorporated to model structural elements and other components. These sheets are available in a variety of thicknesses. There is a wide choice in the visual appearances that can be achieved using these materials since they may be flat, corrugated or perforated to varying degrees of opacity.

Below
Eden Project, Cornwall. Photograph of the completed building.

STEP BY STEP MODELLING A GEODESIC DOME

These images illustrate the construction of a model for Grimshaw Architects' Eden Project, built by Richard Armiger: Network Modelmakers. Considerable skill and planning were required to complete this complex and ambitious model. Once the frames are modelled a model of the landscape is made so that the design can be integrated with its surroundings.

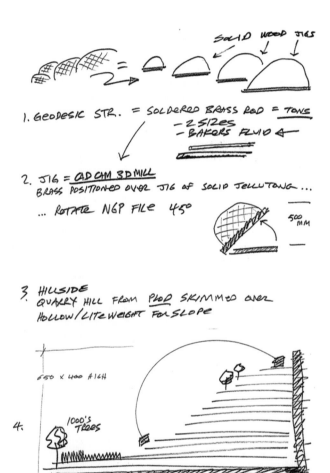

SOLID WOOD JIGS

1. GEODESIC STR. = SOLDERED BRASS ROD = TONS
   - 2 SIZES
   - BAKERS FLUID

2. JIG = CAD CAM 3D MILL
   BRASS POSITIONED OVER JIG OF SOLID JELLUTONG ...
   ... ROTATE NGP FILE 45°
   500 MM

3. HILLSIDE
   QUARRY HILL FROM PLOP SKIMMED OVER
   HOLLOW/LITEWEIGHT FOR SLOPE

650 x 400 HIGH

4. 1000's TREES

5. FOOTING = EXTRUDED & HEATFORM CURVE

1 Initial sketches are made by the Lead Maker to carefully sequence the construction.

2 Brass rod components are cut to various lengths and then taped into position onto a wood former, before being soldered together. The wood former provides the needed support and ensures the parts are precisely positioned.

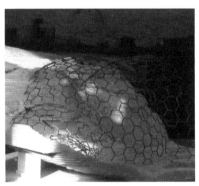

3 To check fit, the sub-assembled geodesic structure is then offered up to the as-yet-incomplete, quarry-hillside site model. This is an essential task because the topography of the surrounding area varies widely.

4 To represent the translucent ETFE panels, the geodesic structure is then clad with a 'skin' of clear acetate components (having first spray-painted the structure with metal etching primer and then added the final colour coat).

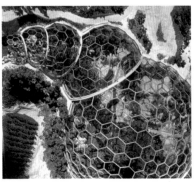

5 To integrate the building with its surroundings, and with the complex geodesic structure and skin nearly complete, the model's hillside context is sculpted, and landscaped with miniature trees and foliage.

Neil Spiller's quest to develop a personal architectural language that investigates the surreal poetics of contemporary technology and identifies new spaces where architecture may exist, has been represented through a wide range of media. His synthesis of mechanisms and myth is illustrated in this sheet-brass model of an intervention for a public space in Piestany, Slovakia.

TIP CUTTING METAL

Cutting metal rods can be difficult so it may be useful to set up a clamp and a block to saw against. This is particularly helpful when making structural frames to ensure that the modular components are of equal length.

The main different types of metal, and their key characteristics, are outlined below:

Aluminium – this is a light silver in colour and does not corrode. It is relatively soft, making it easily worked in models, but whilst it can be glued it cannot be soldered.

Brass – this metal is formed from copper and zinc and is usually seen in its gold format, although the proportions of the metals in the alloy determine the material's colour. It can be polished, soldered and bonded with adhesives.

Copper – this reddish-brown metal can be easily glued, soldered and polished. A unique feature is that it oxidizes through contact with the air, which turns it a green colour over time. Artificially aged or pre-patinated copper is often used for roofs of buildings.

Nickel silver – this is a silver-coloured alloy with a shiny appearance which will not corrode through contact with air. It is useful for representing metallic building components, and can be glued and soldered easily.

Steel – this is typically dark silver in colour and corrodes easily, resulting in a reddish-brown rust. It is a versatile material, which can be welded as well as glued and soldered. Steel can either be protected from corrosion, with paint or galvanization, or it can be weathered prior to use in models to give the appearance of Cor-ten steel. Stainless steel is an alloy that is lighter in colour, shinier in appearance and will not corrode.

When not used in sheet form, the most common type of metal found in models is either wire-mesh or slender metal rods – the latter available, similarly to metal sheets, in different types of metal. Wire mesh is a highly useful modelling material as it can be easily manipulated into complex and organic forms. It can also have additional materials applied to it or stretched over it to give it a skin or cladding. Metal rods, meanwhile, can be cut to represent structural elements such as columns and beams, or may be soldered or glued together to create frames and trusses.

This 1:1,000 competition model for the Mutua Tower, Barcelona, by Rogers Stirk Harbour + Partners utilizes a striking, etched-brass vertical element alongside wooden and acrylic components. The scale and drama of the image is the result of using a wide-angle lens, with the camera positioned as low and as close to the model as possible to communicate a pedestrian-level perspective view.

TIP SOLDERING METAL

Soldering metal components together can be tricky for the inexperienced, so it is useful to practise with scrap bits and off-cuts of material before working with parts intended for a model.

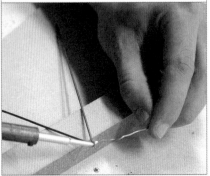

CAD/CAM

The variety of computational modelmaking tools and processes is already very extensive and with the emergence of more and more software and hardware, this field of design generation and representation continues to grow. To discuss the large number of different approaches and techniques comprehensively would require an entire book in its own right, therefore this section describes some of the more common opportunities presented by using computers as part of the modelmaking process. There is no doubt that computers have revolutionized architectural design, and yet despite initial speculation that physical models may become extinct and be replaced by their virtual counterparts, the current situation illustrates that they are in fact experiencing something of a renaissance. In addition, when physical models are produced in conjunction with computerized processes they may present even more advantages for the designer than before.

The most common application of Computer-aided Drawing (CAD) software to produce physical models is through the Computerized Numerical Control process (usually abbreviated to CNC). In essence CNC milling uses CAD software to cut and engrave into the model material as determined by the computer drawings prepared for it. Working in a similar manner to a digital drawing plotter, a CNC machine 'plots' the three-dimensional information from a CAD drawing along three axes. The third axis describes the depth of the 'plot' and may result in the material being cut or engraved. The benefits of this method are the very high degree of accuracy that can be achieved and the complexity of geometry that can be cut. Of course, this process all relies on accurate CAD drawings in the first instance so considerable care must be taken when producing them prior to sending the data to the CNC machine. Once the CAD information has been constructed it can be used as often as required, for example, with different sheet materials fed into the CNC machine or for repetitive model components. Consequently this process affords graphic presentation possibilities on models that were not previously obtainable. Despite these advantages the results are so precise that it is typically used in presentation models rather than for design development work as such models tend to appear too final.

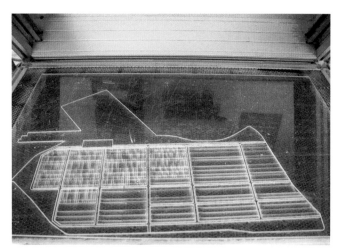

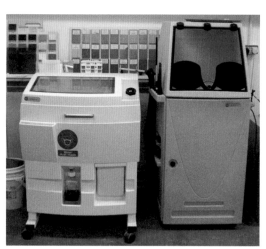

Left
CNC milling machine, used to cut and engrave sheet materials from CAD data.

Below left
An example of an acrylic sheet that has been CNC cut and etched.

Below
Rapid Prototyping equipment. To the left is the machine that gradually pastes layer upon layer of starch powder to produce a physical representation of the digital model. The machine to the right is subsequently used to carefully remove any excess powder and debris from the starch model prior to 'fixing' it with a spray.

Right and below right

The fluidity of circulation and space inherent in the design for the Mercedes-Benz Museum by UNStudio is explicit in this model, made of a resin material built up from layers using Rapid Prototyping. Resin can be used manually, but care must be taken when working with it since the fumes can be toxic. Therefore, always use it in a well-ventilated room and wear a mask approved for chemicals.

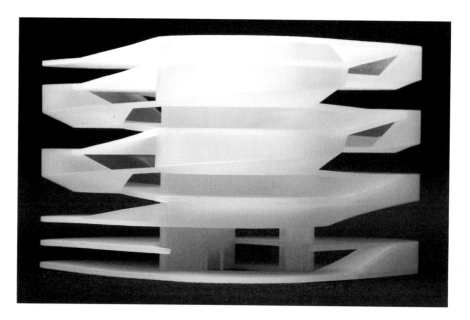

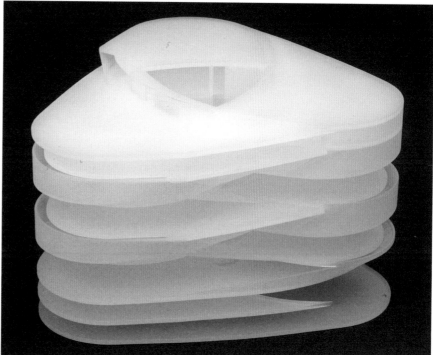

TIP CNC MILLING MACHINES

CNC milling machines can speed up the modelling process enormously, particularly when reproducing façades with many apertures. However, the accuracy of the drawing is vital as this will be digitally converted – so any mistakes, no matter how minor, will be visible. Remember, when drafting out in CAD that lines must always be cleanly and precisely connected at their end points, double lines cannot be used, and lines should be assigned colours to enable the milling machine to define the cutting side of them.

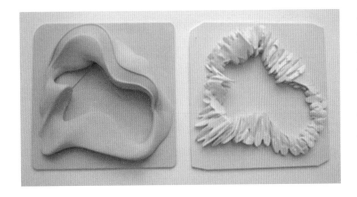

Top

Two models produced using Rapid Prototyping to investigate the spatial impact, boundaries and geometry of experimental forms for a building.

Above

One of the models is shown placed in context. The opportunities afforded by such a process for making complex organic forms mean it can be a very effective method of developing designs and then testing the physical qualities and spatial effects of a proposal. Note that the surrounding urban context is deliberately simplified and made from material similar in tone to the proposal, in order to allow the full drama and sculptural aspects of the design to be appreciated.

A significant number of universities now make the aforementioned modelling processes available to architecture students as part of their workshop resources. As stated earlier, the success of these processes is determined by the accuracy with which the initial modelling data is constructed using CAD software. Therefore, it is essential that all relevant information – such as site topography, façade profiles and apertures – is included before the CNC machine is involved. Without this, the machine cannot operate properly and so extra care should be taken to check dimensions and the CAD file format, otherwise the data will not be correctly 'plotted'. It is recommended that test runs are carried out, as machines may vary – for example, the cutting head may move directly along or to either side of the CAD line – so that valuable time and material is not wasted. Materials that may be used with a CNC machine are those that are available in sheet format, the thicknesses of which vary depending on the material. Metals such as aluminium, brass and steel can be used in thicknesses up to 5 mm (3/8 inch), while acrylic glass and some plastics along with wood-based products such as MDF and plywood can be used in much larger thicknesses depending on the size of the machine. If you are in any doubt as to the suitability of a material, it is important to consult an expert such as a member of the workshop staff.

One of the significant advantages of using computers in the early stages of the design-development process is that it is possible to create and modify forms and structures that would be very difficult to achieve with physical models – especially designs based on complex organic geometry. However, if CAD alone had provided an effective substitute for physical scale models, there would be no need for models made using Computer-aided Drafting/Computer-aided Manufacturing (CAD/CAM) technology – and yet they are produced in an increasingly widespread manner. CAD/CAM modelling can be a full-size operation – for instance, the making of prototype parts from a given material, which is common in a number of design and engineering disciplines. Architects, by virtue of the size of their buildings, typically use this technology to model scaled-down representations of their designs. Some CAD/CAM processes, such as those using a laser or router, are subtractive, in that they remove material from a sheet or block to leave the relevant elements behind – and in this sense, they are essentially a carving technique. By contrast, other processes are additive and involve the incremental build-up of layers of material. An example of this latter process is a Z-starch model that is formed topographically. Firstly, a paste is injected accurately, layer upon layer, into a container of white powder starch. The model's components, such as walls, are the result of thin strands of pasted starch built cumulatively from layers. The paste is then allowed to set, and all the excess starch is carefully blown away to reveal the model. This needs to be fixed with a spray, otherwise the starch simply powders away when handled. Significant advantages of this process for modelling are that it affords a vast range of forms and can make complex interiors with ease. There are of course limitations and, unlike traditional physical ones, CAD/CAM models are very fragile and cannot be revised once they have been made. An additional constraint is set by the machine's assembly box. The maximum dimensions of the assembly box largely determine a CAD/CAM model's size and, as a result, they are not always made to a conventional scale. This reveals an important point about such modelling, in that compared to other media it is expensive and university workshops are currently unlikely to accommodate the huge machines found in industry that are capable of significantly bigger, full-size modelmaking.

Below left and bottom left
Greg Lynn has long embraced and developed innovative applications for CAD/CAM technologies to represent his organic and geometrically sophisticated designs, as illustrated in these models. His Binary Large Object-derived designs have even become shorthand for a particular aesthetic known as BLOB-architecture.

Below
In this structural model for Greg Lynn's proposal for the Eyebeam Institute, New York, the various model components have been laser-cut and then assembled together. A design such as this, with very intricate curvilinear geometry, would be very difficult to achieve using traditional modelmaking methods alone.

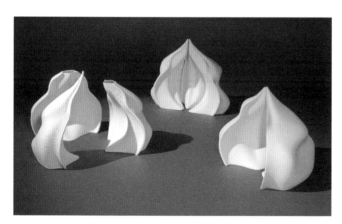

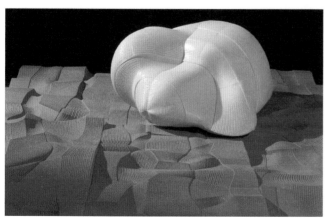

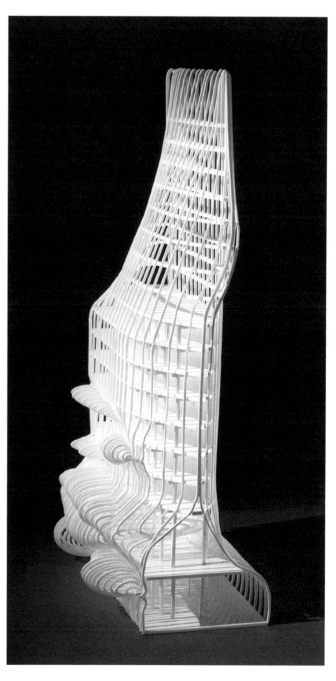

STEP BY STEP DEVELOPING A MODEL USING CAD/CAM

Computational modelling has significantly elaborated the range of generative methodologies and techniques of representation available to the modelmaker. The integration of traditional and digital design tools allows the designer greater ability to experiment than ever before and indeed the gap between different modes of inquiry may be a source of innovation and further knowledge. One of the significant advantages of this synthesis is the exploration of complex geometries and structures, as illustrated here.

1 The first stage of this process requires accurate CAD drawings to be produced.

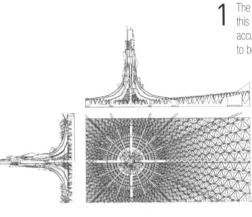

2 CAD data is then used to build a digital 3D model, which may be inspected for its formal properties prior to its physical representation through a CAD/CAM process.

3 The physical model sits in the Rapid Prototyping machine after all digital information has been translated.

4 The model is then carefully cleaned, using a fine airbrush, to remove excess starch powder before it is 'fixed' using a suitable spray in order to reduce wear through handling.

5 The CAD/CAM model can then be used and handled in a similar manner to models made from other media.

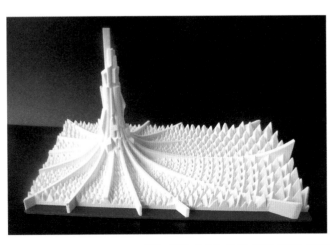

6 The precision of such processes facilitates the manufacture of very detailed components and surfaces, which would be almost impossible using traditional hand techniques.

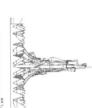

Photography and film

Our visual experience of space and form relies upon our optical contact with the real world. Therefore, when a two-dimensional image is compared with a physical model, the latter – by displaying actual, visual, physical space and depth cues – appears closer to reality. Additional information collected through our other senses, such as touch and smell, enhance our relationship with a three-dimensional object. Indeed, it is the physicality of the model and our ability to move around, touch and handle it that designers typically see as the primary advantage when using its form to communicate their ideas to others. Bearing all this in mind, why photograph a model and potentially lose so many of these advantages of its physical, three-dimensional presence?

The photographing of a model is a transformation into another medium that in some ways is even more critical than the physical construction itself. Our familiarity with two-dimensional images means we have a tendency to accept them as 'real', whether

Above
By controlling the artificial lighting and framing key views, a model may be photographed to produce an array of exciting and atmospheric images.

Right
The dramatic spatial effects communicated in this image were achieved using a comparatively simple model, but carefully arranging the lighting and position of the camera.

via the media screen or printed page. Therefore, a well-conceived and carefully taken photograph can enhance a model and translate it into a more 'realistic' experience. In addition, sequential photographs can reveal characteristics of the design beyond those offered in the holistic overview that a physical model might communicate to the eye. Considered in this manner, the modelmaker becomes a film director, controlling and editing the views of the model and its spaces. The use of different lighting techniques, camera angles and framing permits specific images to be taken, and can have a powerful influence on how the project is communicated to others. Such processes do, of course, come with some responsibility – as images of a model can deceive and seduce as much as they can communicate design ideas clearly. When used properly, photography can facilitate a coherent understanding of a project, and care should be taken to ensure that images that allow for interpretation do not deliberately mislead.

One significant application of photography is its ability to produce large images of what may be a comparatively small model. Models are usually seen from below eye level, and the use of photography enables the modelmaker to orchestrate how a model is seen and may enhance the perception of it. Through the production of large-format images the model is translated into something more visually 'realistic', as issues of scale and miniaturization are rendered less problematic. Further information can then be added to the image – either by hand or via computer techniques, such as collage and

A model-photography workshop using a digital projector to cast colour and shapes onto a model's spaces. The designer can experiment with various camera positions to record the most evocative images.

montage – enriching the composition as desired. Early model photography encountered significant problems when it attempted to overcome depth-of-field issues in model images, but this has since been resolved through technical developments in the medium.

Photographing models is an important part of documenting the design process through which a project has evolved. In both a professional and educational context, such images can be included in a portfolio or further manipulated as an element in photomontages and collages, or combined with CAD information. Therefore, it can often be useful to photograph the various stages of making a model, which will reveal the design development and will also provide images of different elements as they are constructed – for example, the primary structure. The angle or viewpoint of the camera can play a crucial role in convincing the viewer that the scale of a small model is closer to reality. Photographs taken directly above a model should be avoided unless it is a city or urban model – as this is a bird's-eye perspective, through which humans do not usually perceive buildings. The availability of digital cameras means that many photographs can be taken, and the results instantly evaluated and deleted where necessary. Therefore, the photographer should take the opportunity to get a range of viewpoints and close-up shots that can be used to communicate the different qualities of the design. Remember that, although there may be a required definitive image of the design, it is unlikely that all the building's characteristics can be conveyed in a single photograph.

Further communication possibilities are achievable through the application of film or slide projection. The use of projection technology brings a new dimension to the experience of viewing a model – time. Moving images depicting events, changes in colour and lighting conditions, etc, extend the spatial properties of a model and may be used to reinforce concepts behind the design. In addition to furnishing photographs with even more information than those using a model in isolation, these techniques afford the modelmaker the capacity to experiment with a range of temporal design possibilities without physically altering a model.

Lighting a simple cardboard model with two highly contrasting light sources (see tip below) allows the viewer to easily distinguish different moods and spatial characteristics within a design proposal.

TIP INEXPENSIVE LIGHTING

Expensive photographic and lighting equipment are not necessarily required in order to produce creative and atmospheric results. Try experimenting with household lamps, bicycle lights and torches, as these can make great substitutes and facilitate easy manipulation, even in the most confined of spaces.

STEP BY STEP CREATING A REALISTIC PHOTOMONTAGE

Photomontages may be highly effective representational tools as they enable the designer to incorporate large amounts of information and atmosphere, and communicate this to the viewer. The careful positioning of the camera when setting up this type of image provides a more realistic view of the building and needs to be from a similar perspective as the site context, making the montage of these images easier and more convincing. Further 'lighting' may then be added during the photomontage stage to provide a more coherent visual relationship between the design proposal and its surroundings.

1 A physical model is produced as carefully and accurately as possible. This is particularly important where images are to be enlarged, as even the smallest imperfection will be considerably magnified when the photograph is blown up.

2 The same model is placed in a darkened environment and an image is projected onto it, recreating the effect of illumination across the façade.

3 The 'realistic' photomontage of this image is the result of modelmaking, good photography and subsequent blending with contextual information using digital software.

A Single Lens Reflex (SLR) camera will typically produce good results, although the effectiveness of its images will be partly determined by the scale of a model and how accessible it is. The widespread use of digital photography can be a real asset to the modelmaker, as it enables many different views and lighting conditions to be tested without wasting valuable time and printing resources. This process also grants the modelmaker the ability to carefully choreograph exactly which shots will be the most impressive, and to communicate the essential aspects of the design. Often, an image of a model alone may not provide enough information for the viewer, and, indeed, may reinforce the notion of it as an isolated object or work of art in its own right. Consequently, photographs of models are frequently superimposed onto an image of the site or context, enabling the viewer to further appreciate the design's relationship with its surroundings. If such images are to be produced successfully, then care should be taken to photograph shots of both the model and its context from similar viewpoints, as this will make their integration easier and much more convincing. Modelmakers will also photograph their models against the sky or a simulated sky backdrop to add to the illusion of reality.

TIP BACKGROUNDS

Always think carefully about the view of the model that you wish to capture. Neutral backgrounds make a model the focus of the viewer's attention, and make it much easier to test different lighting conditions upon and within it as required. Avoid having any objects in the background that will instantly reveal a model's real scale.

Below left
The use of a simulated sky background in this image enables the model to have a more integrated relationship to its context, even though the site is represented in a fairly basic manner.

Left
This photomontage places the model as a relatively small element within the overall composition of the image, meaning that the viewer's attention is directed to the scale and form of the design proposal in its context rather than any detailed information on the model.

Above
The key to the success of this image lies in the careful location of its viewpoint. Although it is effectively a bird's-eye view, both the model and its immediate surroundings have been photographed from very similar positions, thus allowing the combination of the two images to be seductively accurate.

Digital and camera technology

This particular type of media raises an important issue about models and their function. A model, as emphasized earlier in this book, is an abstraction of reality to a variable degree depending on its purpose. One of the most immediate ways in which such abstraction is evident is in the scale of a model. The distance created between an architectural model, and the awareness of the viewer's own size in relation to it, is known as the 'Gulliver Gap'. In the past, one of the ways in which architects counteracted this phenomenon was to make models large enough to afford interior views at eye level. With the developments of camera, film and digital technology, coupled with the considerable expense in producing such models, this practice is far less common nowadays. A very useful piece of equipment for negotiating the scale barrier is the endoscope, also known as a modelscope, which brings the eye directly into the spaces within a model. Working as a miniature periscope, an endoscope can be directly inserted into a model, affording views of the interior spatial sequence that would not be possible from looking at the whole model. Attaching a digital camera to the endoscope and saving the resultant images to a computer can enable these views to be photographed and further manipulated where necessary. Similarly, moving images and space-time relationships may be explored and communicated by attaching a digital video camera to the endoscope and then editing the film as required.

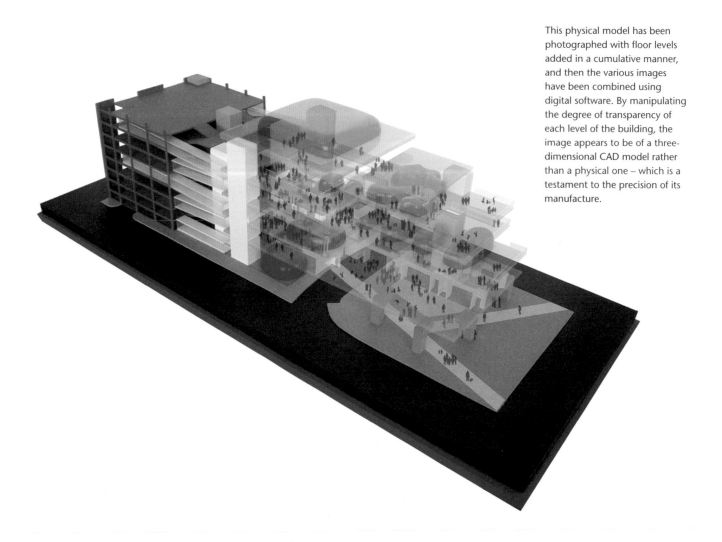

This physical model has been photographed with floor levels added in a cumulative manner, and then the various images have been combined using digital software. By manipulating the degree of transparency of each level of the building, the image appears to be of a three-dimensional CAD model rather than a physical one – which is a testament to the precision of its manufacture.

Below
Endoscope images from within
a physical model enable the
sequence around the events spaces
of a building to be shown, and
can be used to communicate the
spatial narrative and characteristics
effectively. In this example
miniature-scale figures have been
incorporated into the model,
but this type of image could be
manipulated using digital software
to add further definition to spaces
as desired.

Below
1:100 model and endoscope.
The mesh envelope was deliberately
made freestanding so that it
could be removed while using the
endoscope camera, facilitating
easier use – remember not to fix all
model components together until
all the images required from within
a model have been recorded!

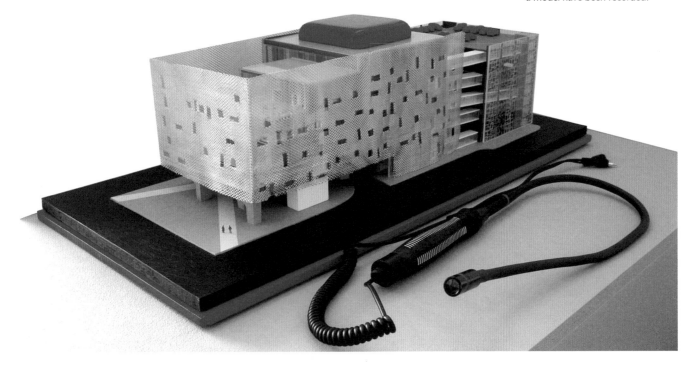

TYPES

Introduction

In the previous section, we looked at a broad range of materials and at suitable techniques for using them in architectural models. While it may be tempting to pick up the basic tools and immediately start the process of making a model, it is worth considering one's objectives first. Making a model can be a highly creative way of designing a building, and a pleasurable activity in its own right, but the purpose of the model should be identified from the outset in order to maximize the results given the limitations on time, materials and effort. In the first instance, it is useful to establish what the model is to represent or generate in terms of ideas. That is not to say that the outcomes will definitely be known at this stage – particularly in the case of design-development or explorative models – but that a basic set of goals are borne in mind, which should be achieved through that model's construction. These will automatically have an influence on the choice of media, scale and degree of abstraction within the model. This therefore leads us on to the topic of different types of models, as it is very difficult to embody all possible aspects of a design in a single model alone, especially during the early stages of a design process when the end result is unknown.

Architects face such decisions on a regular basis, and when presenting their ideas to clients or professional committees have to consider what aspects of the design need to be communicated to their target audience. As a general rule, people teaching architecture have more familiarity and experience with assessing designs and will appreciate abstract models as 'statements of architectural intent' more readily, perhaps, than members of the public. The opposite of abstract is concrete. In painting, 'concrete' refers to a portrayal of an object that is as accurate as possible. Within the context of architectural models, abstraction shifts the focus onto the subject matter of the design, the informational value of the object portrayed and its spatial framework. At stake is not an accurate portrayal of reality but a process of simplification, which guides the eye to the model's essential features. It is crucial to find a suitable form of abstraction, one that reflects the selected scale. At the start of any project, modellers must select the scale of the model they intend to make. The level of detail with which variously sized models can represent an architectural object illustrate the role that scale plays in the modelling process. Depending on the scale, and the level of abstraction required, there exist a number of model types, which will be explained below.

This concept model was made using strips of card, mesh and metal to represent the different degrees of enclosure and flow around an urban site. Its sculptural qualities result in an evocative abstraction of materials as it captures an initial response to site conditions.

Concept models

A major development in the academic environment of architecture has been the growth of theory within the discipline, and the resultant increase in the use of conceptual models. During the twentieth century, and, more specifically, throughout the last 40 years the subject of architecture has undergone a significant transformation, both in the nature of debates within it and in its relationship with other academic disciplines. This shift can be emphasized as described by Neil Leach in his introduction to *Rethinking Architecture*: 'not only are architects and architectural theorists becoming more and more receptive to the whole domain of cultural theory, but cultural theorists, philosophers, sociologists and many others are now to be found increasingly engaged with questions of architecture and the built environment'. (Leach, N., *Rethinking Architecture*, London: Routledge,1997, p.vii) One of the practices that runs parallel to this emergent field is the proliferation of models as generative and representative tools of conceptual ideas in architecture. The use of conceptual models as a medium of thought not only facilitates the design process of the modelmaker but also enables his initial creative impulses or intentions to be communicated to others.

Above
Printed portraits across the surface of this thin card model reinforce the community aspect of the design for the CPlex centre, West Bromwich, by Alsop, whilst still retaining an abstract quality that, in formal terms, is merely suggestive.

Below
The emergence of a concept can lead to innovative models being made out of whatever may be close to hand. In this example, a peeled aubergine with its skin repositioned in strips using pins makes a provocative representation of initial design ideas about a façade wrapping around the building core.

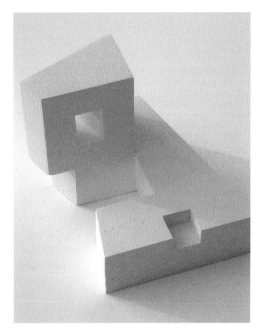

Philosophy has long used architectural terminology as a source of metaphor, but the growth of theoretical discourse and its dissemination through books and journals has seen an emergence of both critical theory and contemporary architectural theory across architecture schools around the world. A key consequence of this development is that a student's design often can begin with a theoretical position or concept that is not necessarily related to the functional programme of the building, or even its context, but is evidence of a novel idea. This type of inquiry is often expressed using a variety of media including, but not exclusive to, painting, sketches, text, computer visualizations, audio-visual recordings and models.

The use of a conceptual model as a device to initiate an architecture student's design process in response to a project brief is common practice since it is a very useful tool of communication, mediating as it does between the theoretical ideas in a student's head and the 'concrete facts' of architecture such as structure and functional requirements.

A diagram can generally be described as conceptual, as it distils the information being communicated down to its most essential. In modelmaking, the underlying idea of a design or a creative concept is depicted in an entirely abstract manner as a three-dimensional object – often at a metaphorical level. Material, form and colour highlight structures and create compositions. The model can, for instance, be used to visualize the results of urban space analyses at the start of the design process. By exploring a theme or place on a spatial yet abstract level, architects can alter or improve the view of that place. A model can support this approach and, owing to the high degree of abstraction typical in this type of model, can incorporate novel objects as part of its composition. One particular feature unique to these types of model is that they are not necessarily made to scale, being representative of initial creative impulses that do not seek to communicate actual spatial relationships.

Top

In their competition winning entry for the Leventis Art Gallery in Cyprus, Feilden Clegg Bradley Studios used two sculptural concept models to communicate formal ideas about the building. The first model represented the massing of the building in cast and polished plaster of Paris, a material deliberately chosen to be heavy and tactile.

Above

As a counterpoint, the second model represented the three key spaces of the building as green Perspex jewels held within a minimal wire frame.

Site/city models

This type of model represents urban or natural environments. Following on from the concept model, it is the first step in the actual representation process since it shows the design's relationship with the existing environment. In terms of urban space, it is important to show how the context changes with the addition of a new structure.

This type of model is often characterized by one of the highest levels of abstraction. Buildings are reduced to 'building blocks' – abstract structures that reproduce built form and three-dimensionality in a highly simplified manner. Even so, depending on its scale the model may include characteristic features of buildings such as recesses, projections and roof designs. With urban contexts, it is often useful to make a model at 1:2,000 or 1:1,250, as these easily correspond to map scales and allow cross-referencing of further information. In its abstract form, the site – the scaled-down landscape – is simplified and depicted, in the chosen material, as a level plain. Where a landscape slopes, it can be broken down into horizontal layers that are stacked on top of each other in the model.

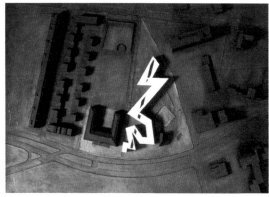

Top left

This model by Meixner Schlüter Wendt for the Ordnungsamt (Public Affairs Office), Frankfurt am Main, typifies the use of a single material to represent the urban context and maximizes the effect of the proposed design, which is made from a variety of coloured components.

Top right

Daniel Libeskind's iconic Jewish Museum in Berlin was in part developed from a component of a stretched Star of David. The 'lightning bolt' qualities of its plan are further underscored here by the use of a contrasting clean and light material against a darker and more brooding context.

Bottom

Jewish Museum, Berlin. Aerial photograph of the completed building.

The development of a project might result in several designs, and urban design models are often constructed as 'inserts' or group models to reduce the amount of work required to present them. In the environment of architecture education, it can be useful to have only one model of the surrounding area made and each student given a mounting board on which to model the portion on which he or she is working. This particular portion is omitted from the urban design model, so that the inserts can be interchanged.

Right

This example uses the above technique to enable different design ideas for the towers and corresponding public space to be tested without the need to make the complete model over and over again. Clearly, because the adjacent surroundings will impact upon – and may, in turn, be affected by – each design proposal, these inserts are made as a removable element of the model (as indicated by the use of a darker material).

Below right

This model uses the contrast and interplay between mute, coloured, solid elements for the existing urban condition and transparent forms with bright, multi-coloured details to highlight various features of the scheme.

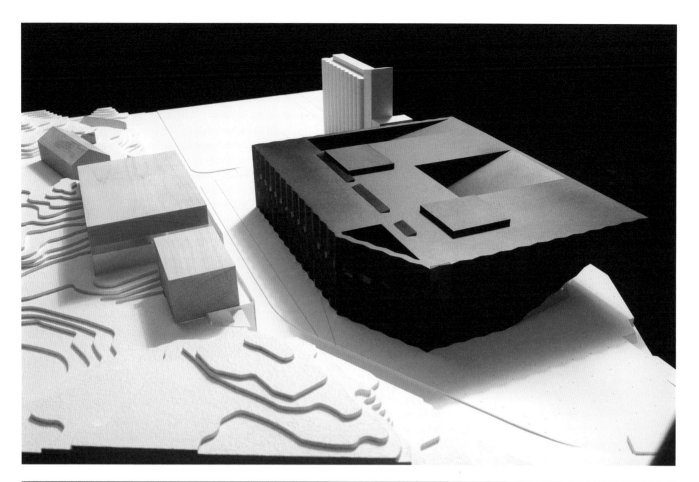

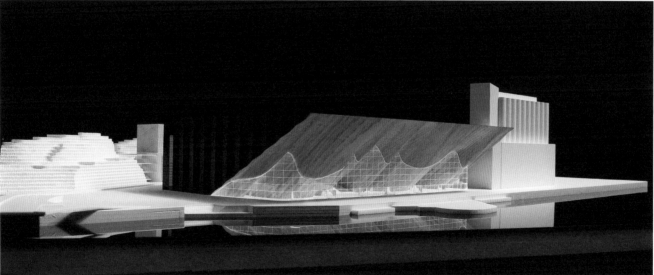

The materiality of the building in this example is reinforced by its relationship with the adjacent white context and high-gloss base which, in the bottom image, is used to represent the water of the nearby river. Viewed from this 'ground-level' position, the project is much more effective than when seen from above (as shown in the top image).

In situations in which a model is required to represent uneven landscapes and topographical data, the first step in building it is to conceive of the irregular natural terrain as a stack of horizontal strata. The more finely layered the material used for this, the more precise and homogenous the resulting model will be. The work will be based on a map or plan that shows the contour lines, or that at least provides topographical information in elevation. Once the real topographic situation is known, contour lines (curved, straight or polygonal) are drawn. Depending on the material, the modeller can cut out each layer with a knife or saw before arranging the strata on top of each other.

Left

This series of models by team-bau proposes urban design solutions across a range of scales. Central to the concept is the use of existing infrastructure and networks within the city, which are communicated as separate layers stacked above satellite images at the appropriate scale.

Above

In this model, the natural topography of the landscape is an intrinsic element of this proposal for a house which offers several strata of accommodation, including a swimming pool sunk into the hillside. The use of the same material to make both the building and the landscape offers a cohesive integration of the scheme within its surroundings, yet also enables the viewer to easily distinguish between the two elements.

Above
The reflective, translucent 'leaves' that trail across this model clearly indicate the new architectural proposal in relation to a monochromatic urban context. The implied movement of the dynamic new architecture reinforces the concept of a climatic 'vortex' that was used to generate the initial ideas.

Below left
This model for the School of Art & Art History, Iowa City, by Steven Holl Architects uses bold colours to identify the key features of natural landscape, water and building. Based upon notions of layering, the architecture consists of planes that seem to float and extend into the immediate surroundings rather than communicating the appearance of a 'closed' object.

Below right
School of Art & Art History, Iowa City. A photograph of the completed building.

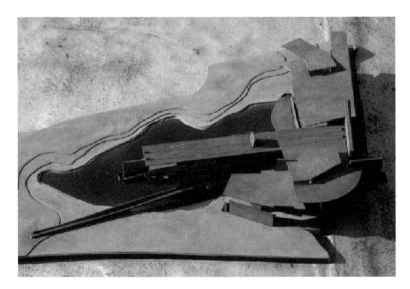

Block/massing models

Brightly coloured forms communicate the different functions and areas within this playful model, which was used to investigate the relationships between internal elements both as a visual composition and as elements in the requirements of the building's programme.

This type of model is similar to the site and city models described above, in so much as it provides a simplified communication of a design's various components rather than detailed information. Depending on the extent of the design, the various components may represent the different spaces of a single building or a number of different buildings that form a complex. The main distinction is that they communicate the relationship of that building's elements or its whole only in relation to itself rather than the surrounding context. This enables a designer to investigate the formal qualities of a design – such as proportion, shape and mass – without necessarily becoming too entrenched in more specific issues related to materials, construction and detailing. Such models are very useful tools as they enable architects and students to make quick design decisions and test any novel ideas they have, as well as providing a refining process for initial thoughts. This type of model is often used by students to examine how different parts of a building's programme may stack up and be rearranged in relation to both aesthetic and pragmatic considerations – an example of the latter being proximity to other buildings and circulation.

Below

In the very early stages of the design process for this building, models were produced in order to create designs in three dimensions without any drawings. This series of sketch models rapidly explored possible massing options using styrofoam and cut, clear acrylic to represent cores, atria, etc.

Bottom

Rather than using blocks to represent volumes, this model utilizes planes to define the primary spaces within the scheme so that their interrelation and impact on the outline of the urban context may be observed.

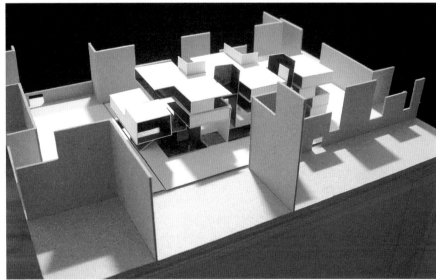

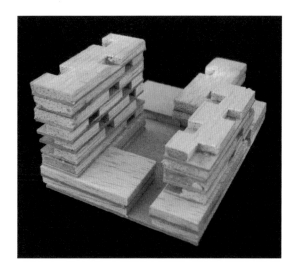

Left
A massing model, exploring the composition of various layers within a project design and the stacked effect of its voids.

Below
The full impact of this design for a public building is instantly established by its bold colour and size in relation to the low density of its surroundings.

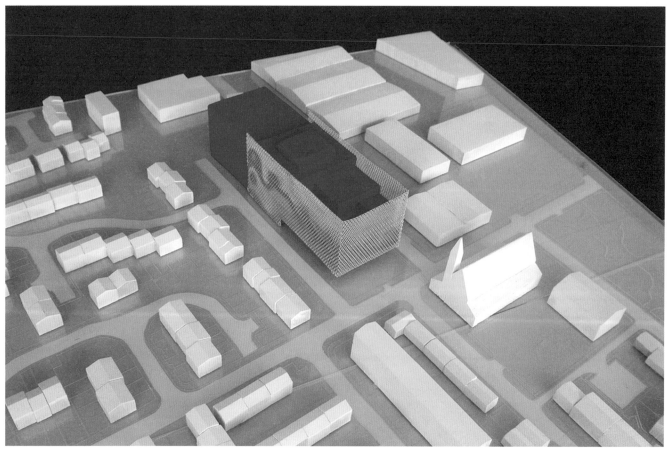

Another advantage of these models is that they can be made using a variety of different media, so that while they may not necessarily attempt to replicate final building materials, the effect of colour, light and mass can be explored both between the various components and within the overall composition. A significant number of designers use this technique to investigate different qualities of various materials in relation to initial design ideas – a process clearly evident in the proliferation of physical models currently being produced in contemporary practice.

Left and below

Alsop Architects have an established portfolio of innovative, playful and vibrant projects. Integral to their design process is the development of ideas through the production of many colourful models, made from a variety of materials. These all have a dual function, being used both for exploration in design terms and also providing vital tools in client/public participation and engagement events.

Design development models

Design development or 'process' models are effectively three-dimensional sketches through which novel ideas are explored and tested but not necessarily concluded. The relationship between design development models and final presentation models is not always clear. Not all presentation models are the result of a sequence of design development models; in some cases a presentation model is a design development model that has reached a critical point in the design process. A key characteristic of design development models is that they are always made by the designer, unlike other types of model that may be subcontracted to professional modelmakers. Perhaps most significantly, design development models communicate a 'journey' rather than a 'destination', as they explicitly illustrate the thought, effort and time committed to investigating design ideas. They often represent the evolution of a design and showcase the sometimes trial-and-error nature of the creative process. This is an important point, as it is tempting to assume that architects arrive at design solutions with a 'lightning bolt' of inspiration and can immediately visualize this in a coherent and convincing manner.

The majority of models produced in architecture schools are design development models, made to explore possibilities and reach a suitable response to a studio project brief. Whilst these types of model may

This 1:1,000-scale design development model for the Leadenhall Building by Rogers Stirk Harbour + Partners was rapidly produced by simply drawing on paper and then folding. It is a good example of how effective and quick sketch models can be. Compare this to the 1:200 presentation model of the same project, shown adjacent. Although this model was the first to show the building with any level of detail – and has been widely used by the client in presentations and exhibitions throughout the planning process – its relationship to the earlier model is clear.

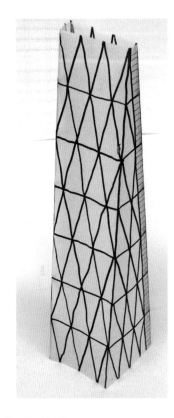

appear unfinished and unrelated to one another, they can be grouped together to indicate the sequential development of design ideas as part of a review critique. When presented alongside drawings and other forms of visualization, these models serve a dual function. On the one hand, they reveal the design evolution of a project, while on the other communicating a student's methodology and learning. Typically in schools of architecture the final set of drawings and model are rarely assessed in isolation, and this means that some documentation of the design process is key to the evaluation procedure.

As stated above, this type of model is part of the design process and functions as a tool enabling the designer to express emerging ideas. Therefore they are quickly produced, often from a range of materials that may be to hand. The preciousness of the idea is the overriding factor in such models, not the perfect presentation of it. Indeed, the temporality of design ideas is visible in these models as a variety of possibilities are explored. For example, Frank Gehry times the production of his design development models at around 3.4 minutes each! (Ivy, R., 'Frank Gehry: Plain Talk with a Master', in *Architectural Record*, vol 5, 1999, p.189)

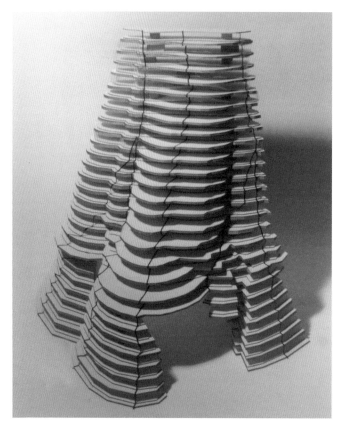

Above
This model was made as part of a project to design an urban block that could maximize its exposure to natural daylight. Laser cutting the plates prior to assembly meant that the innovative form of the 'block' could be easily and quickly produced.

Below
Design development models may encompass various scales, depending on the stage of the design process during which they are made. They may investigate ideas in relation to context, the building itself and internal relationships – or more detailed parts thereof.

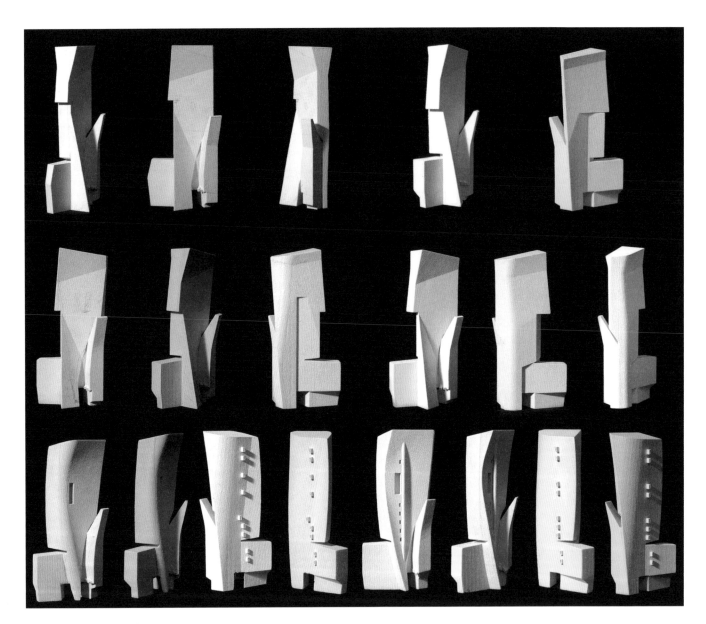

The pursuit of a design solution as a response to a project brief results in the need to continually create, evaluate and revise different ideas. Sometimes this process can result in the further revision of a previous model, or a novel idea may emerge through the simple act of handling materials in three dimensions. Morphosis' design for the Phare Tower, Paris, was rigorously pursued through design-development models at all stages of the design. The evolution of the building's sleek, curved form is clearly evident, as subtle variations in its volumes, geometry and detail are made and examined.

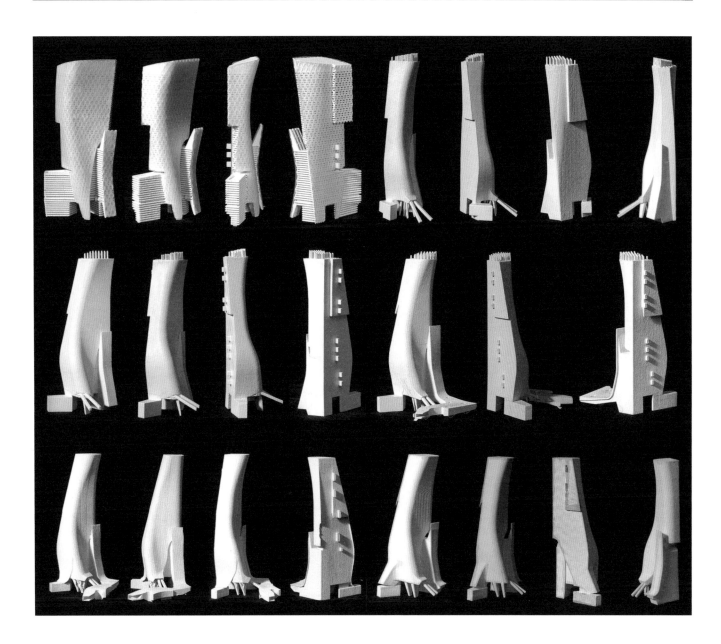

Design-development models do not necessarily represent a complete building, nor do they need to be accurate or at a consistent scale. They may reveal latent characteristics of a design, such as circulation flows or other geometrical relationships. A shift in scale can enable this type of model to be a flexible tool, as some ideas become more defined and others dissolve away. The immediacy and explorative nature of design-development models means that they can be reinterpreted, recycled and even collaged together for future projects as they embody the very spirit of creativity.

The building model is a common illustrative tool for simulating an architectural design, and typical scales for models are 1:200, 1:100 and 1:50. Larger schemes, however, are usually represented on a scale of 1:200, and 1:500 is common in architectural competitions. In addition to three-dimensional forms and volumes,

the design's many and diverse features play a more significant role here than in urban design models, as more consideration may be concentrated on various aspects. Façade design is extremely important, and may even lead to specific models that are solely concerned with investigating the interface between interior space and exterior 'place'. The building model can also convey such information on interior space and structure. For example, a sectional model can provide views of important rooms. In combination with removable floors and other interior components, a modeller may also use a removable roof element that affords a glimpse into the model from above.

STEP BY STEP MAKING A WAX MODEL

The experimental nature of developing a design idea can lead to new avenues of exploration, which will often provide unforeseen and exciting results. Models can be a very useful medium through which to both generate and then further explore ideas that may be subsequently applied across a range of scales. Considered as modes of inquiry, models facilitate the pursuit of novel ideas that may in turn lead to design revisions and refinement. In this example, wax is used to investigate mass, volume and light.

1 A mould is created by pouring rubber around a wooden positive model.

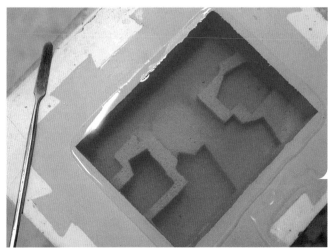

2 The mould is then filled with melted paraffin wax.

3 The cooled wax positive model is carefully removed from its mould.

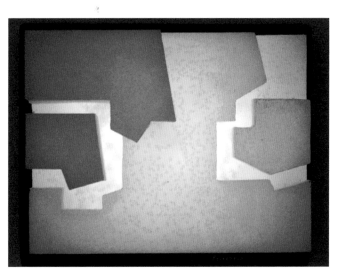

4 The wax model is then evaluated under different lighting conditions, or backlit as in this image, to investigate the resultant effects and materiality.

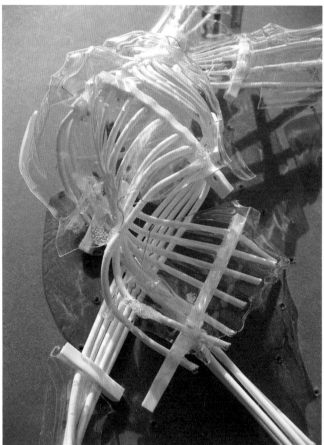

Left

The process of design development frequently lends itself to a series of models that afford the designer to test, edit and reconfigure ideas quickly and effectively. This type of model may often be produced in a reduced palette to allow design variations in relation to form, volume and light to be readily perceived. In this example, the development of the Imperial War Museum North by Daniel Libeskind further explores the relationship between the main building and landscape elements and illustrates a number of revisions from the original competition model (see page 134), most notably the removal of a link bridge across the canal.

Above left and above

The development of a design often provides designers with the opportunity to experiment with new techniques and processes. In this example a skeletal frame has been made as a concept model to evoke the fluid geometry of the initial idea (above left). This is then photographed and used to develop architectural ideas. In this instance the modelmaker has used the process of vacuum forming, also known as vacuforming, which is a simplified type of thermoforming to mould a thin sheet of plastic around the skeletal model. After this, various parts of the mould were cut away to form spaces and degrees of enclosure for further investigation (above).

Case study Process models

The use of process models to explore design
possibilities is key to the design philosophy of
OMA's Rem Koolhaas. So significant are design-
development models to the practice that they
are often used to communicate their designs to
clients, external organizations and the public.
In this series of images, a model for the Casa
da Música building in Porto is manipulated to
reveal the relationships between the design's
overall mass, inner volumes and voids. The final
photograph shows the completed building.

Spatial models

An important difference between a descriptive presentation model and an explorative model is that the former seeks to provide a holistic view of the finished project, whilst the latter may be produced to investigate particular components of the design. As the form of a new design emerges, a whole series of more specialized building models may be constructed that respond to questions arising from the initial evolution of the architectural form. These are specific spatial-study models, i.e. complete or part models specifically built to explore certain issues. By working directly in space, albeit at a small scale, concepts are formed by a student and a design is refined as a result of its exploration in three dimensions – a process in which options remain open in design routes, which might not be as obvious to the designer solely using two-dimensional drawing methods. Spatial models, therefore, are characterized by the fact that they may only focus on the relevant attributes of a design that are critical to a particular space or sequence of spaces. In this sense, they may appear similar to interior architecture models but are typically made at an earlier stage in the design process when internal qualities are less defined. Perhaps most obviously, spatial models do not usually communicate external envelopes or façades as their primary function is to explore the internal arrangement and composition of a building's programme – and in this way they are more quantitative than qualitative in their investigation. They may represent specific and subtle explorations that would be very difficult to visualize within a holistic building model, and may allow design ideas to be interrogated in a rigorous manner.

In this model, the spatial relationship between possible circulation routes around the design's internal volumes is investigated.

Below and bottom
Grafton Architects use spatial models to examine and develop the internal characteristics of their projects. This process of interrogation leads to the production of a series of models for each scheme, as illustrated below. Subtle variations are combined with radical revisions in the quest for the optimum design.

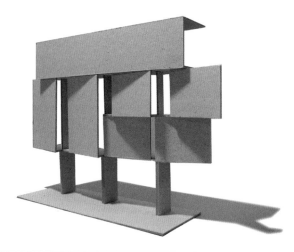

Left
The practice also employs spatial models to analyse the spatial effects of a façade design, here simplified to the essential elements using one material.

Below
Spatial models are not the sole preserve of architects, and may be used to visualize the effects of objects within existing spaces. This example is a 1:100 model for a work called *Hatch* by the artist Antony Gormley. The concept is that the experience of the room should feel both energizing and dangerous, a sense of concentration of space that has been provisionally limited by a room but that extends beyond it. This model was made to help visualize the total effect of the sculptures within the space.

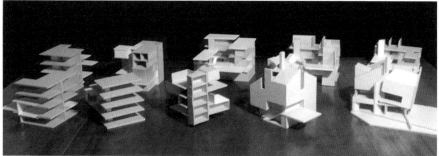

The spatial complexity of Peter Eisenman's geometric explorations is evident in these models for the Guardiola House, Cadiz. By having the components mounted on acrylic sheet, the spaces appear to float, connected only by a ribbon of circulation winding through them. The fragmentary nature of the design is revealed as the viewer moves around the model with each movement reconfiguring and reframing space, as illustrated in these two distinct views.

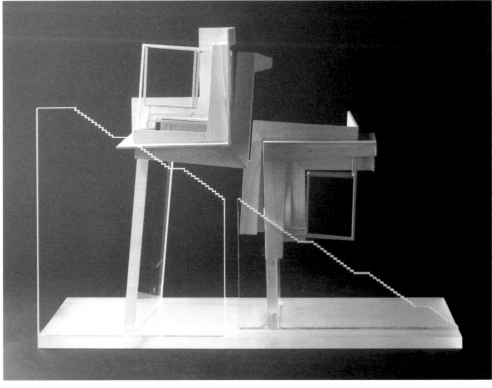

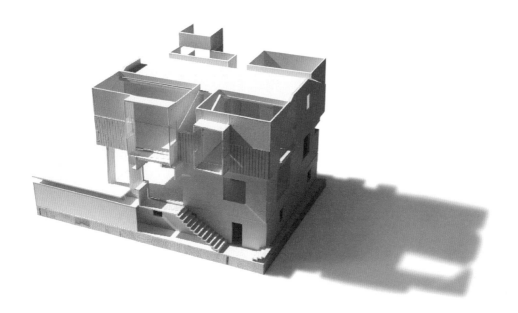

Left

A model made to investigate the spatial qualities of a house's volume and apertures. The deliberate use of a monochromatic material allows the full interplay of light and shadow to be studied.

Below

By projecting lighting onto this model, the effects of its penetration across and beyond the design's spaces can be observed. This technique can be a useful method of investigating how an architectural proposal may appear in the evening and at night.

Below and bottom right

Spatial models may not necessarily be preoccupied with the design and communication of the actual, physical characteristics of a building. In this example, a set of models has been made to explore the circulation systems of a design and how these relate when stacked together.

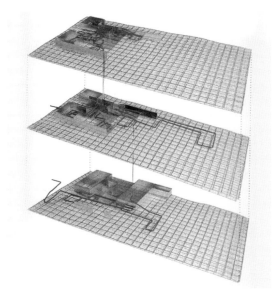

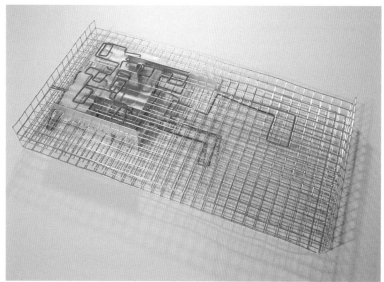

Structural models

As the name suggests, these models are primarily concerned with testing ideas for a project's structure. These are very useful models from an educational point of view, as they enable both a student and a tutor to gauge how well the student understands the complex process of construction. This type of model varies from basic models, that may simply indicate different loadbearing elements such as walls and frames, to considerably more detailed examples that investigate the structural systems of a design with a higher degree of architectural analysis. A particular advantage of these models when compared to drawings is that they allow the designer to understand how various components support each other and fit together in three dimensions – an issue that can often be difficult to comprehend in two-dimensional representations. Furthermore, they facilitate the testing of novel structural concepts and design ideas. It is important to engage with this type of model during the design process, as initial concepts and spatial explorations should always be reinforced by the structural rationale behind the design rather than compromised by it.

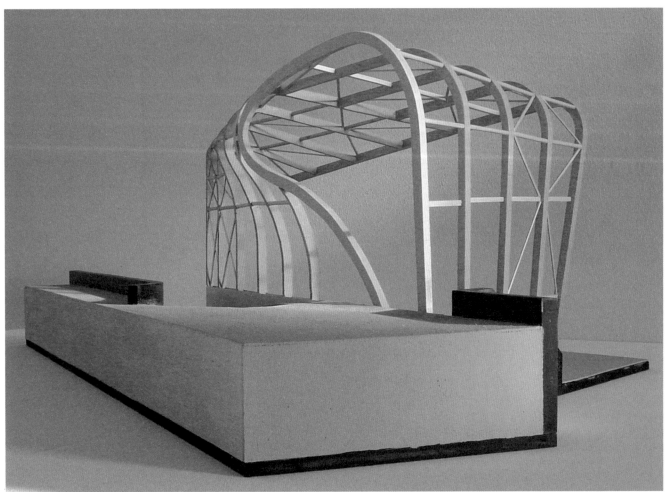

Opposite top
This model has been made to investigate the hybrid structure of loadbearing and frame elements for a student's design proposal to test if it will work.

Opposite
A structural model for the Architecture Centre Amsterdam (ARCAM) by René van Zuuk, made to examine the primary structural elements of the curvilinear envelope. For more images of this project see pages 72 and 124.

Above and right
Archi-Tectonics' design for the Gipsy Trail House, New York, was developed through the use of an integrated structure which connected to its immediate environment. The overriding concept was to create an intelligent structure and to design the house 'from the inside out'. The morphing of the building's programmatic elements resulted in a segmented, organic shape. This example illustrates how structural concepts may be used to drive the design development of a project.

In terms of structural intentions, physical models may be more immediate in identifying potential problems than CAD models, which may appear deceptively straightforward as they exist in a 'weightless' space unbound by gravity. A feature of the majority of structural models is their skeletal appearance, since their function is to represent the primary components of a structure and their assembly. This results in many of the other building components being left out, so that a clear understanding and communication of the structural strategies can be achieved.

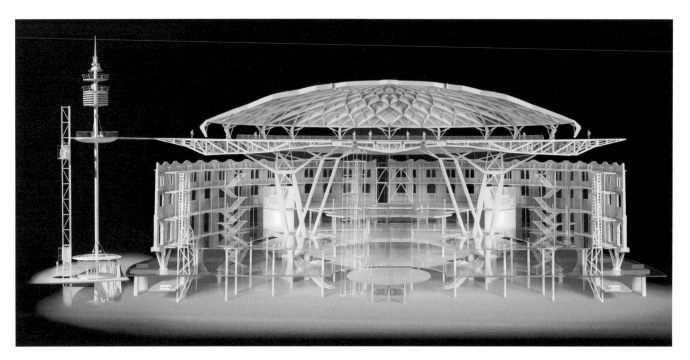

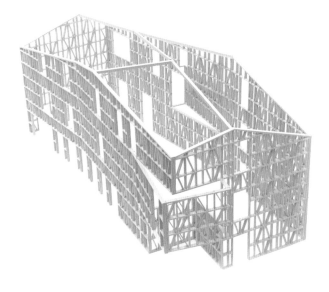

Left
A model illustrating the structural framing required in the external envelope of a project.

Above
This 1:50 model for the scheme at Las Arenas, Barcelona, by Rogers Stirk Harbour + Partners actually shows a quarter model mirrored to show the full section. The large model was used to explore the complex geometry of the new intervention into the existing bullring structure. It was used as a working model in the truest sense, in that many design elements were tested within it. For the build phase the model was taken to site, where it has been invaluable in conveying the design to co-architects and contractors.

Right
This sectional model demonstrates the contrast between an intricate, curvilinear canopy and the rectilinear spaces below. Models such as these can be an effective method of communicating the variety of structural ideas within a design without representing the entire building.

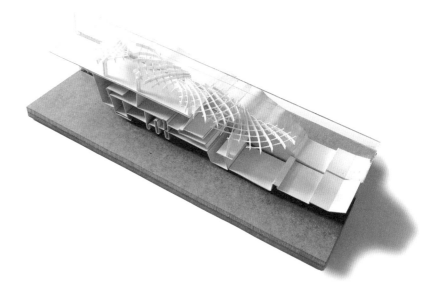

Left
A laser-cut model illustrating the elaborate cross-sections of the new roof structure for the Santa Caterina Market, Barcelona, by Enric Miralles – Benedetta Tagliabue/EMBT. In this example, a structural model has not been produced to test a design proposal but as a communication tool to enable the viewer to understand the complex geometry of an existing piece of architecture as a precedent.

Below left
This model was made using rapid prototyping to analyze the loadbearing structure of an existing nineteenth-century Victorian Gothic building prior to any new intervention being designed for it.

Below right
A skeletal model for Peter Eisenman's design of the Max Reinhardt Haus, Berlin. The building assumes a prismatic character, folding in and opening out on itself as a reference to the constantly changing array of metropolitan experiences and relationships. The development of the form is generated from a single Möbius strip whose vectors are extended into plates, the extremities of which are connected in a triangulation of their surfaces, as illustrated here.

Interior architecture models

Interior architecture or design is widely recognized as a discipline in its own right, and there are many published precedents and much information available on the subject.

Depending on the scale of the design proposal, it can frequently be useful for architects to make models of interior space for both design development and communication purposes. Such models are solely produced to explore the internal characteristics of space(s); as such, they are not necessarily interested in the overall design but focus on a particular part of it rather than the composite whole. As a result they may be quite crudely finished externally, as it is their interior qualities that are important. The need to investigate a space in detail usually leads to these models being built at 1:20, 1:10 or 1:5 scales. This type of model typically incorporates detailed components such as staircases, furniture and miniature people, but care must be taken as the level of detail and scale of the models can easily resemble a doll's house. This potential problem, as with other models, is reliant on the appropriate degree of abstraction. In other types of models with less detail this is usually easy to overcome, but in interior architecture models, where the intention is to provide an accurate and detailed simulation of the real internal environment, it is not so straightforward. Consequently, the level of abstraction is reduced in this particular type of model; its main focus is to communicate materials and objects on a small scale.

An interior architecture model used to evaluate the reception area of a project. In addition to the analysis of daylight conditions, this example also permits the designer to examine the effects of the proposed artificial lighting scheme.

Right

A sectional model such as this enables the architect to explore the relationships between the internal spaces and the external skin of a building. This is particularly useful here, since the internal spaces are 'hung' within the overall volume of the design, and the distances and lighting conditions between interior and outer envelope vary considerably.

Centre and below

Interior architecture models may be useful tools for examining the relationship of objects within a building. These internal views of the Mercedes-Benz Museum by UNStudio allow the designers to examine the role of the circulation in relation to the exhibited cars. Note also the use of photographs at the back of the model to illustrate context and potential views out of the building.

In this type of model, the most effective way of representing materials is to use the actual material itself, but care should be exercised to ensure that such application does not affect the illusion of scale. When made with care and skill, it is possible to achieve superb effects with interior architecture models. They can be selectively photographed, making it difficult for the viewer to discriminate between a modelled space and the real situation. Interior architecture models can be wall mounted, permitting the viewer to look into, as opposed to onto, the contained space. The advantage of this technique is that it improves our perception of the spaces: the viewers feel as if they are really 'there'.

Above left
The dramatic effect of a large glazing element, and how it illuminates the forms within a design, can be studied using a model such as this example, for ARCAM (Amsterdam Centre for Architecture) by René van Zuuk.

Above right
ARCAM, Amsterdam. Photograph of the interior of the completed building. For more images of this project see pages 72 and 118.

Left
The stacked floors of this model are placed adjacent to a projected image of the context in order to communicate the design's extensive views and open-plan characteristics. Furniture and scaled figures are placed to create a more realistic image and to prevent the model appearing abstract and without scale.

Below
A view looking inside a model from the exterior. This example is notable for the use of figures behind translucent material on the first floor in order to cast silhouettes and describe activity. The careful framing of the image in relation to the perspective of the model draws the eye into its space.

Case study Exploring light and shadow

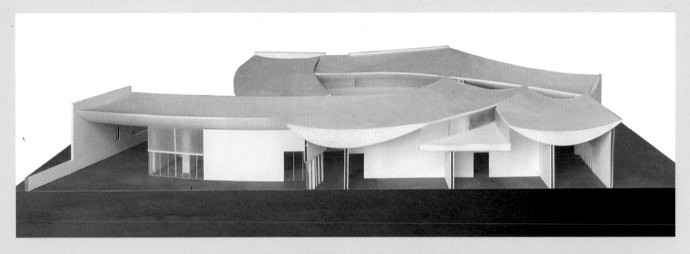

The architecture of Steven Holl frequently delights in the careful interplay of light and shadow upon his considered composition of forms. This approach has led to numerous commissions for gallery designs – such as the Herning Centre of the Arts in Denmark, featured here. Conceived as a fusion of landscape and architecture, the gallery spaces are orthogonal and simple with good proportions in respect to the art, while overhead the curved roof sections bring in natural light. Design development of the gallery spaces was explored using a sectional model, shown in the top image. Photographs of the internal spaces and the effects of daylight within them were recorded and people were added using digital software in order to provide a sense of scale, as illustrated above left and right. An essential aspect of most gallery designs is the control of daylight to ensure the exhibited artworks are not adversely affected. Typical positions for paintings were subsequently added in order to communicate the pragmatic qualities of the design, as demonstrated in the centre right image.

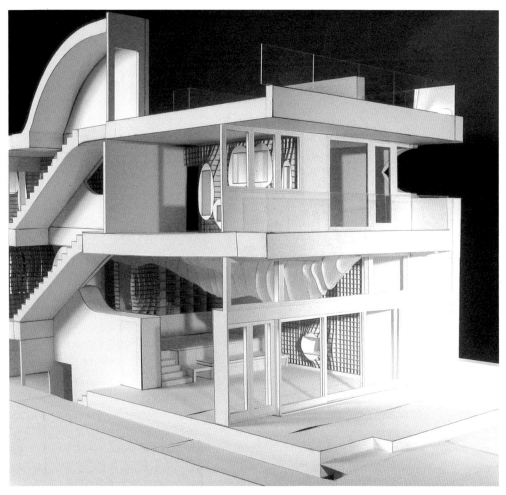

Left

The dynamic between curvilinear and rectilinear geometry in the interior of this house design can be fully appreciated by the designer through the use of this sectional model. A typical characteristic of interior-architecture models is that they seek to represent true spatial relationships, and so whilst a degree of abstraction is necessary, owing to their scale they enable the viewer to look inside the spaces and observe how they may actually appear in reality.

Below left

A model of an auditorium space, made to study the effect of the curved roof upon the volume. Note how the image has been framed to enable the viewer to look into the model's space rather than be drawn to the edges of it, which would detract from its illusive properties.

Below right

This model was made to permit the designers to examine the space between the double skins of the proposed building.

This model for the Slavin House, Venice, California, by Greg Lynn enabled the designer to inspect the lighting effects of the apertures upon the voids within the building, which were a key feature of the design.

Lighting models

Lighting models may be used to produce both qualitative and quantitative data. On the one hand, they may allow the designer to perceive the effects of daylight or artificial light throughout the spaces modelled; on the other, they may be used to record more accurate information in relation to various amounts or combinations of light sources and the subsequent effects these have on the model's spaces. They may also make strategic use of light in order to project an image of the proposal that seeks to explain more latent or poetic qualities within the design. This type of model can create an impressive effect by incorporating miniature bulbs, fibre optics and transparent or translucent materials. Such models are often used to emphasize specific characteristics of a design. Beyond providing dramatic and atmospheric effects to a project, lighting models can be used to effectively communicate particular designs depending on their function. Buildings typically have a different appearance at night than during the daytime, and a lighting model can convey such implications of a design in its context.

Lighting models do not necessarily represent realistic spatial properties, and some models have lighting incorporated into them in order to add drama and atmosphere to the design – or to emphasize the formal qualities of a project.

For Feilden Clegg Bradley Studio's design for the new Manchester Metropolitan University Business School and Student Hub, this model illustrates the overall 'lodged jewel' form of the building along with its internal organization. The envelope of the model conveys some of the ideas of incorporating dichroic glass fins into the façade and roof forms. The projecting fins would not only protect the glazed areas of the building from unwanted solar gains but also, with their varied colour refractions, give the different faces of the building a distinct character. In order to make the removable envelope of the model robust enough to be handled, the fins were 'inverted' and carved out within the thickness of the Perspex envelope by laser cutting the internal faces to various depths. These inverted fins glow by reflecting coloured light, which is transmitted up through the clear acrylic envelope from hidden light-emitting diodes (LEDs) in the base of the model.

STEP BY STEP INVESTIGATING LIGHT EFFECTS ON AN INTERIOR

The use of lighting models permits designers to try out lots of different ideas by simple variations in direction of view, colour and level of illumination. This is often an important aspect of such modelling since it enables the designer to make further revisions as a result of this process and also acts as a powerful communication tool to illustrate to others the potential qualities of the proposed spaces. In this example, the designer sought to investigate the effects of the surrounding city upon the design's interior space.

1 It is often useful to sketch out ideas prior to testing them, as this collage illustrates. The central vertical space of the design has structural trusses to its left, and the urban context lies beyond.

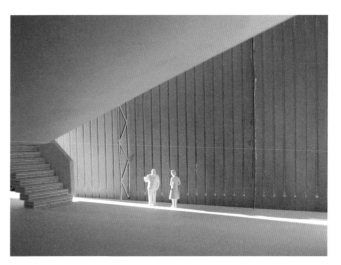

2 A model of the space to be examined is constructed. Here it is shown with a directional light projected from the side, which may give some expression of how the space will appear in daylight.

3 By projecting a multicoloured image onto the back of the translucent wall of the model, it is possible for the designer to evaluate how the atmosphere of the proposed building may change at night.

Below and below centre
Notice the transformation of the minimal cardboard model shown in typical daylight conditions in the upper image, and as it is illuminated from within to represent how it may appear in the evening.

Below right
Lighting models may not simply represent an entire building but may comprise more detailed studies of specific spaces – such as this example, made to evaluate the lighting within an elevator interior in relation to the atrium beyond.

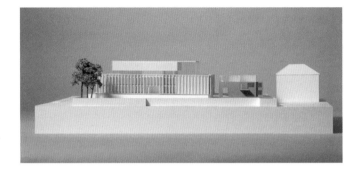

TIP USING LIGHTBOXES

The use of a lightbox underneath a model can show a design in a completely new way. In this example, the relative voids and mass of the scheme become less significant than the public space they enclose – allowing the designer to investigate this aspect in further detail.

Above and above right

In this house design by Scape Architects, a dramatic acrylic wall and staircase was conceived as a unifying element within the scheme. The translucent wall, which extends the full height of the building, casts a complex play of light and shadow throughout the house, filtering light down through it. A model was therefore produced to analyze the qualities of light around this key feature. Note the difference between the images of the model, since the one to the left simply describes the physical characteristics of the design but the right-hand image is top-lit, affording evaluation by the designer.

Right

The use of blue light in this interior architecture model allows the designer to view spaces in a new manner, as the daylight of the design studio environment may not be the same as the daylight conditions at the proposal's location. The sculptural qualities of the design are also highlighted by this application of light, and thus it may also serve as a presentation model.

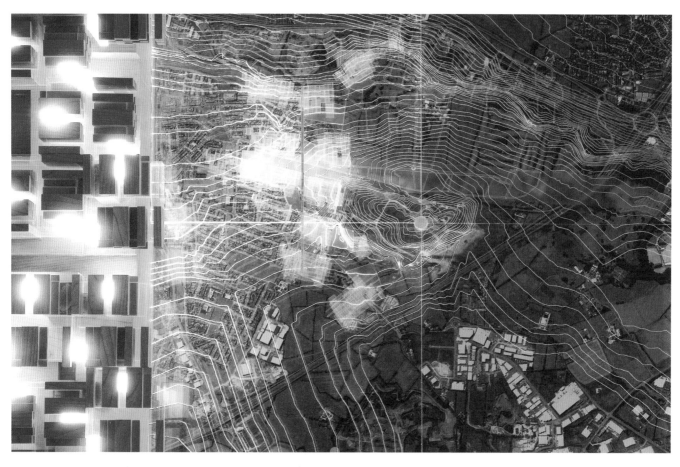

By virtue of lighting this model from underneath the relationship between the planning proposal, indicated by the blocks, and its location, shown by the illuminated contour lines, becomes a cohesive presentation tool. The light is diffused through acrylic blocks in the massing element of the model and via perforations in the context photograph, which has elevated above it a layer of sheet acrylic etched with additional site information.

Presentation/exhibition models

Perhaps the most familiar type of architectural model, due to its frequency in the public eye, is the presentation or exhibition model that typically describes a whole building or project design and signifies a point in the design process at which the designer is ready to communicate the proposal to external audiences such as clients and public. This type of model often has a major role in the communication of a design proposal, as it is often the one subject to most scrutiny. Such a model may also act as a 'talisman' for the project owing to the use of its image in the press, on websites, etc. Why should this be? A presentation model provides a clear and coherent description of a design, and functions as a representation of the building on that building's own three-dimensional and formal terms. However, no matter how precise this type of model may be it still provides critical distance by virtue of its scale and thereby maintains a degree of abstraction. The presentation model affords communication and understanding rather than any real knowledge of the thing itself, i.e. the completed building.

In some instances, models may be built specifically for an exhibition. This is particularly common for examples of historical architecture, for which no model may actually exist or for which an extant historical model may be badly damaged. It also enables the exhibition designer to liaise with the modelmaker and, where relevant, use such models as key elements of the overall exhibition's composition. The presentation of this type of model, whether to a private or public audience, may present opportunities and constraints depending on what needs to be communicated.

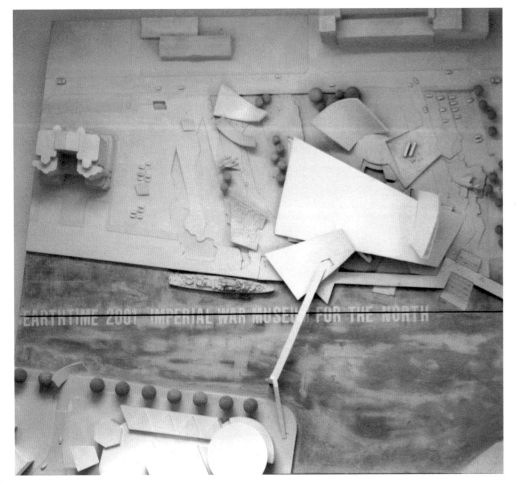

Left
Competition model for Daniel Libeskind's design for the Imperial War Museum North, Salford. The model describes the composition of the building, conceived as fragments from a literally shattered globe.

Opposite
Major retrospectives of architects' work often feature models as their centrepieces, as is shown in this example of an exhibition of projects by Will Alsop and his practice. They afford a great degree of engagement with the audience, and unlike two-dimensional media can immediately communicate design ideas and spatial characteristics with flair. Note the mounting of models on plinths or tables, which is typical as this enables the viewer to appreciate them at eye level.

Below

The role of the model as a powerful communication tool is demonstrated in their prevalence across international design competitions as typified by this competition-winning model for the Singapore Gardens by the Bay design by Grant Associates/ Wilkinson Eyre, made by Network Modelmakers.

Opposite top

This sectional model of The City of Culture of Galicia, Santiago de Compostela, Spain, designed by Peter Eisenman, illustrates the concept of using the city's symbol of a scallop shell to develop surface-like forms. The buildings are inscribed into the ground as architecture and topography merge together, offering a series of carved spaces rather than building blocks.

Opposite centre and bottom

Presentation model for a social-housing scheme in Madrid by Morphosis. The model describes an alternative to typical towering blocks of faceless units, by proposing a radically different social model that integrates landscape and village topologies. A layer of landscape overlaid upon a façade, which is composed of a series of open spaces and idiosyncratic punctures, serves to break down the institutional nature of the public-housing project, as shown in the lattice-like circuitry of this model. A photograph of the completed scheme is shown below the model.

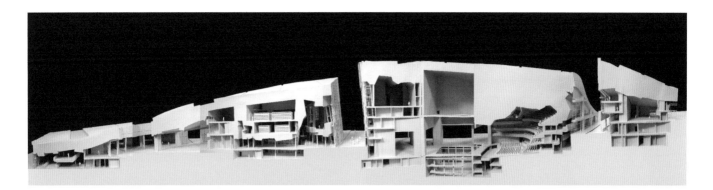

Above left

Peter Eisenman's proposal for the Church of the Year 2000, Rome, presents the idea that modern transportation and the media have effectively collapsed the distances involved in pilgrimages. As a result, the design was developed through different folded iterations to explore ideas of spatial sequence and 'journey'.

Below

It may be useful to make a presentation model in which various elements can be removed to allow more detailed communication of design features and a greater understanding of the interior. Neutelings Riedijk's model for the STUK concert centre in Leuven, Belgium, incorporates removable floor plates allowing the stacking of the building's programme to be clearly understood in relation to the overall scheme.

Above right

This presentation model for the ISTA Lecture Hall, Vienna, was part of Greg Lynn's competition entry. Contrasting materials of perforated brass sheet and wooden veneer emphasize the organic, bubble-like geometry of the new design proposal in relation to the existing built fabric.

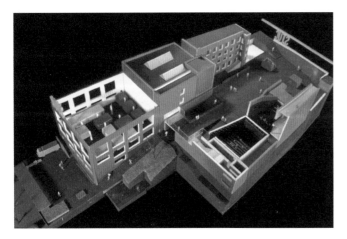

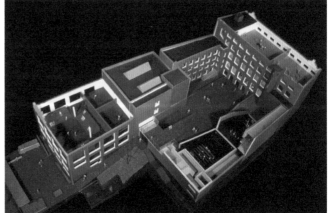

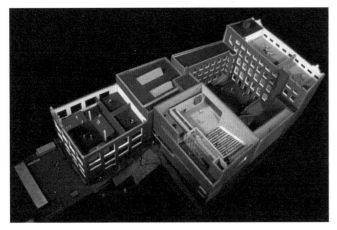

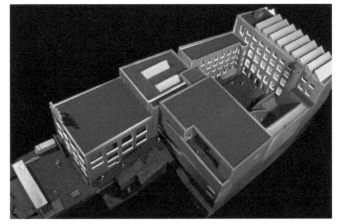

STEP BY STEP MAKING A PRESENTATION MODEL

Many firms outsource their presentation models to commercial modelmaking firms, where versatility, experience and, in recent times, modelmakers' interpretation come to bear. Modelmaking for architects and designers has become an industry in its own right, with the best firms offering high levels of precision, craftsmanship and, especially, speed in order to meet international deadlines. Modelmaker Richard Armiger observes that ambitious presentation models are rarely made in an ad hoc manner, and typically require a great deal of

'strategic planning in order to meet eye-watering deadlines'. On some occasions an awkward building detail is resolved by the modelmaker 'in one of his "interpretive moments''... The modelmaker is indeed part of the design process.' (Interview with Richard Armiger, 11 September, 2008). The sequence shown here illustrates the construction process of the model for Coexistence Tower by Future Systems, built by Richard Armiger: Network Modelmakers.

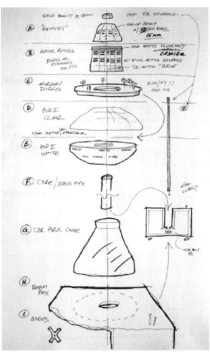

1 Helpful when relaying information to a team of makers, the Lead Maker may prepare quick sketches to describe the preferred materials, recommend a sequence of construction, illustrate different sub-assemblies and so forth.

2 Richard Armiger being assisted by his then student apprentices, John Applegate and Simon Hamnell.

3 Once completed, the model should be documented through photographs. These can be used in photomontages, to record the design's evolution, and especially, to promote the design.

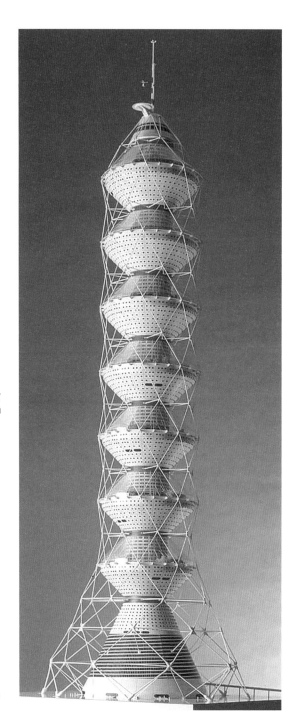

Case study Models for an exhibition

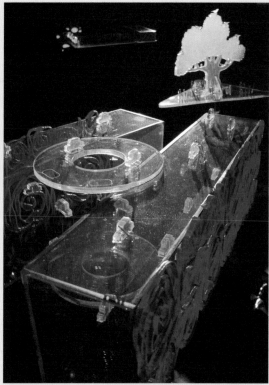

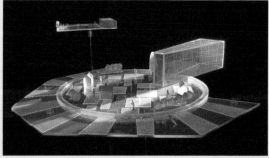

The complete installation, featuring models of the scheme by Hawkins Brown, for the redevelopment of Park Hill, Sheffield, exhibited at the Venice Biennale in 2006.

Exhibitions often require architectural designs to be displayed in interesting ways in order to capture the public's imagination, and this can lead to the development of installations within gallery spaces to provide optimum environments for viewing models.

For their proposal for the regeneration of the Park Hill housing development, Hawkins Brown commissioned a series of coloured acrylic models to illustrate the various community-inspired elements within their proposal. These were then suspended on metal rods, with images hung and projected around them. Combined with atmospheric lighting, as seen opposite, the overall effect of the installation was both dynamic and dramatic.

I LOVE YOU   WILL U MARRY ME

From Object of Despair to Object of Desire: 10 catalysts for change

Full-sized prototypes

Detailed models are not only used in the field of interior design, but also as structural or technical models known as 'details'. In principle, these models can be made to a scale of up to 1:1, in which case it would probably be more accurate to call them 'protoypes'. Twenty years ago, it was not uncommon for architecture practices to invest in full-scale mock-ups of building components, interiors and even entire floors of high-rise projects in order to investigate the design implications involved. This is obviously a costly process and, although it does still occur, the use of CAD has enabled a significant amount of a design's potential characteristics and behaviour over time to be predicted and evaluated electronically. However, this type of model is particularly useful for the designer who may have difficulty understanding how elements combine in three dimensions. Full-size models are often limited by their scale, and as such it is typical to use them as a vehicle to explore detailed components more rigorously rather than to attempt to make a replica of a significant portion of the design proposal. This is not necessarily a practical exercise when time and space are limited, but the benefits can be significant.

This 1:1 sectional mock-up of a modular building enabled the designer to evaluate the construction of the space and then subsequently use it as part of an exhibition to reinforce the design proposal.

The 'twist' in the concrete structure of the
design for the Mercedes-Benz Museum in
Stuttgart involved considerable engineering in
order to ensure stability in a building of such
three-dimensional complexity. This full-size
mock-up of part of the structure was made
so that the architects (UNStudio), engineers
and the construction team could assess its
performance on site.

Case study Prototype models

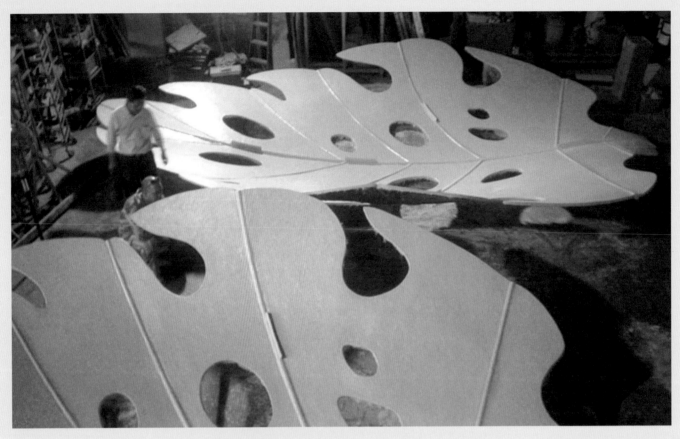

Above
Production prototype canopy for the scheme for One North Park, Singapore.

Below left and right
Design development through prototypes for the Madrid Rio project. The tree support's form was initially made in wood and, when suitably refined, a mock-up was produced in cast iron and tested in situ.

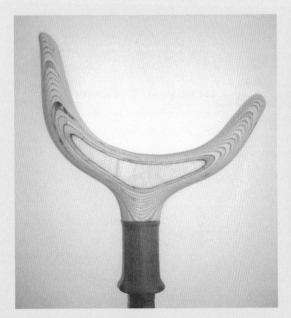

West 8 has applied a multi-disciplinary approach to complex design issues in large-scale urban masterplanning and design, landscape interventions, waterfront projects, parks, squares and gardens. Based upon the knowledge that the contemporary landscape is for the most part artificial and made up of different elements, both designed and undesigned, the practice has responded by positing its own narrative spaces. It identifies the basic ingredients of good public realm as ecology, infrastructure, weather conditions, building programmes and people. Its aim is to incorporate the awareness of these various aspects in a playful, optimistic manner that stimulates the desire to 'conquer', and take possession of, space. A key aspect of the practice's projects is the making of full-size prototypes in order to evaluate the impact of design ideas on the site, and to allow further development and appropriate revisions.

For its proposal for Jubilee Gardens in London, the practice introduced the idea of an undulating landscape with white seating 'cliffs'. Polyester-concrete panels and strips were fabricated especially for the design, with foam-dummy prototypes produced prior to manufacture in the final materials.

Below
This prototype model was made by the designer in order to explore a novel hydraulic structure. The use of rapid-prototyping technology enabled a detailed physical model of considerable three-dimensional complexity to be produced from CAD data rather than through labour-intensive hand techniques.

Material appearance can frequently be deceptive when shown on drawings, and even the most sophisticated textured and coloured computer renderings do not compensate for the loss of tactile information. As a result, architects may use models to explore the potential materiality of building surfaces and components prior to their application within a design. This is particularly important in situations in which the materials may have an unknown visual impact – or, indeed other unforeseen properties.

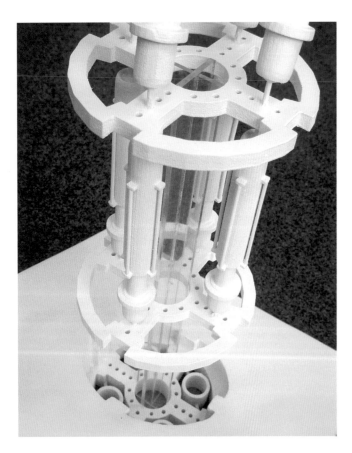

Top right
The materiality of this surface, intended for use in a public space, is produced full-size in order that its visual, tactile and weathering characteristics may be observed on site.

Above
Further detail studies are produced in order to identify the most appropriate level of texture for the relief panels in the composition.

Below

A full-size mock-up of a façade panel, facilitating both the design and construction teams to evaluate the combination of materials on site and to identify any potential assembly difficulties prior to the construction of the final building.

Right

1:1-scale models are not always feasible owing to the time, cost and space required to build them. However, this type of model can be very useful for testing components of an overall design. In this example, a test model for a seating feature is first produced in wood to allow the designers to evaluate its ergonomic properties.

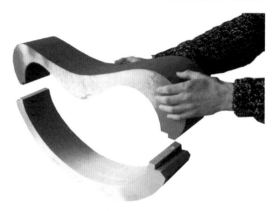

Above

Once the design has been consolidated, a prototype section of the seating is manufactured using the materials intended for the final object – in this case, high-density foam.

STEP BY STEP MAKING A CLADDING MODEL

Evaluation of material surfaces and other building elements is not an activity exclusive to architects. The innovative modelmaker may wish to explore properties of various components for consideration to be included as a feature in a design, which may strengthen or relate to the original concept. In this example, an investigation into the suitability of a textured component for use as a cladding panel is explored through a straightforward process of modelmaking. The tactile nature of the test component enables a designer to get, literally, a 'feel' for materials. Depending on their intended use within a design, elements such as these can be photographed and applied at a range of scales offering further design possibilities.

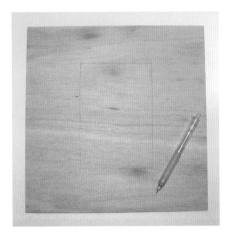

1 The size of the test component is marked out onto a piece of board.

2 Double-sided adhesive tape is then applied in strips across the marked-out area.

3 A material – in this instance, thick string – is laid out across the adhesive tape in the desired configuration.

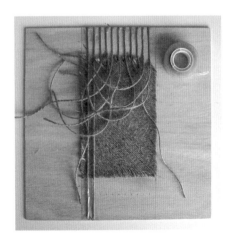

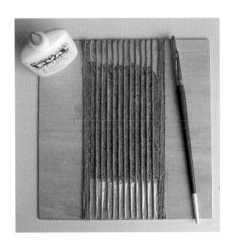

4 More material is taped in a vertical fashion across the first layer of string. This will form a relief pattern in the model panel.

5 PVA glue is applied in a thin layer across the area in order to prevent the material sticking to the plaster as it dries out.

6 A wooden frame is placed over the textured area, and all edges are sealed to prevent liquid plaster leaking out.

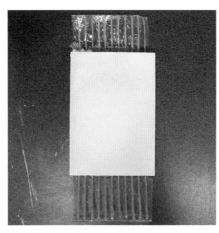

7 Liquid plaster is then poured into the frame and allowed to dry out completely, in accordance with manufacturer's instructions.

8 The frame is then unclamped and removed.

9 The textured layer is then carefully peeled away from the plaster.

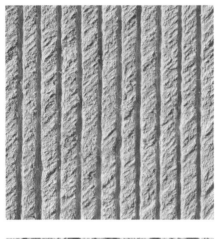

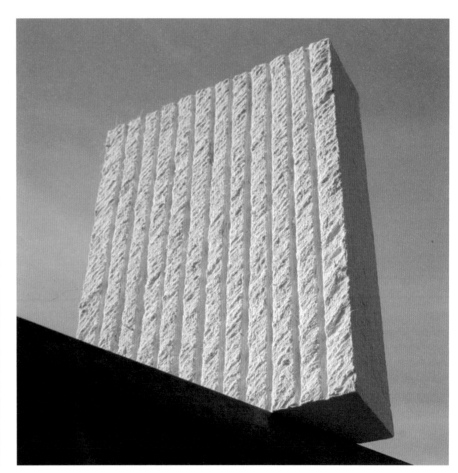

10 A textured component is revealed, which can be studied in various lighting conditions in order to afford visual appreciation.

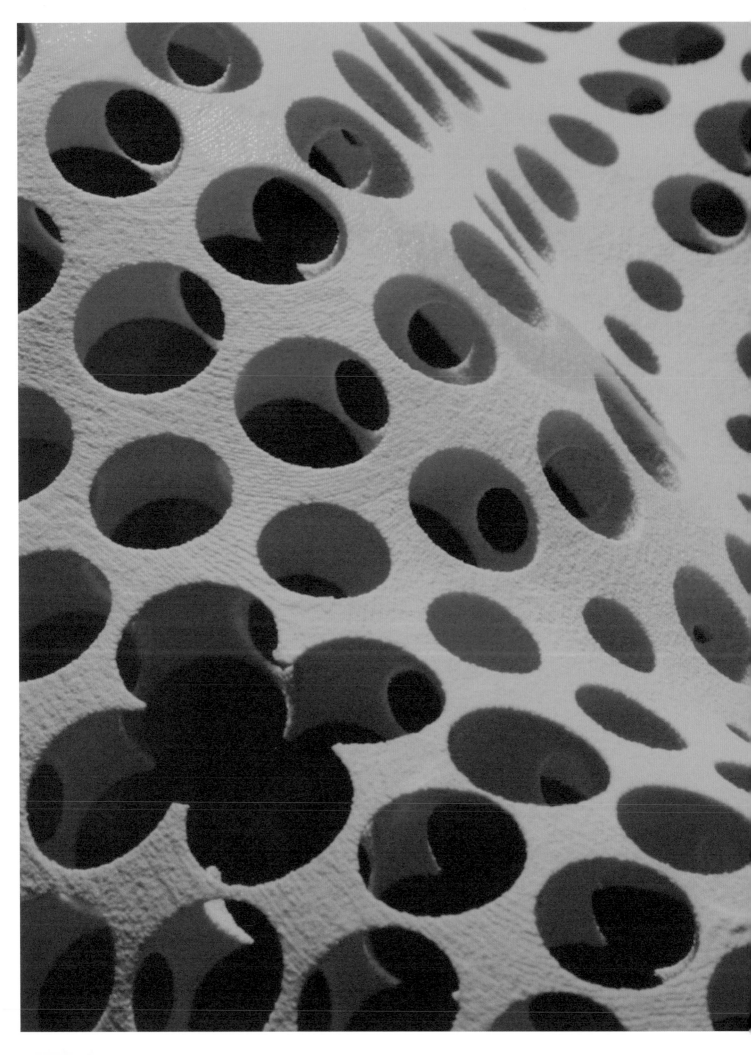

APPLICATION

Introduction

In the previous sections of this book, we have examined the nature of models and why they are such important design and communication tools. We have also looked at the various media from which models can currently be made and how these can be combined, where appropriate, to provide different types of model. Clearly, models may be an integral part of a designer's working practice – and yet they are so common in the exchange and development of ideas that we rarely think to give them a great deal of attention, and they are frequently used without question. At the start of this book, the point was illustrated that architectural history and practice are paralleled by a history of models as diverse in form and function as the buildings and ideas they seek to represent. From highly abstract conceptual pieces to full-size prototypes enabling evaluation of the assemblage of innovative building components, the range of modelmaking opportunities available to the designer is both inspirational and occasionally daunting – particularly to those new to the discipline, or to those learning modelmaking techniques. With this in mind, the final section of this book will attempt to demonstrate and explain not just what a model is made from and its corresponding purpose, i.e. the media and type of model, but when, how and by whom it is used within a design process. The reason for this is to deepen the reader's understanding of models and their application. Perhaps most interestingly, models should be considered as dynamic objects that are not necessarily 'frozen' for a specific point in the process of designing architecture but can be revisited, revised and (re)presented to different audiences. This reinforces the recurring theme that models are both generative and representational in relation to design ideas. In essence, there are four different applications for models: descriptive, predictive, evaluative and explorative.

Above
A descriptive model for Coop Himmelb(l)au's design for the Akron Art Museum, Ohio. The relationship between the dynamic, angled elements of the proposal and the existing building is consolidated and the model is made from materials that echo those used in the completed building, so that the design can be effectively communicated to the client.

Right
Akron Art Museum. Photograph of the completed building.

Descriptive models

If we are to define the descriptive model in general terms, then its purpose is to assist the understanding of reality by establishing the emergence of a particular phenomenon and describing relationships between relevant factors. Put more succinctly, its primary intention is explanatory. As has been mentioned earlier, any description of reality brings with it many issues regarding accuracy. The most familiar type of descriptive model in architecture is, perhaps, the 'presentation model'. The presentation or descriptive model typically illustrates the complete and fully detailed arrangement of an architectural solution, often within the context of its immediate surroundings. A key characteristic of the presentation model is that it usually signifies the end of a critical stage – and even, sometimes, the culmination – of the design process. Distinguishable from the exploratory architectural model used for design development, this application of a model incorporates miniature components of the architecture represented

in detail and as a complete, finished entity. A significant difference between the decorative representation of a building and the representation of architecture needs to be made at this point. Descriptive models typically make the architecture coherent and easy to understand; for example, such a model may feature cladding panels or other such construction and structural components that are arranged to clearly indicate how the finished building is intended to look. By contrast, if we refer to the example of a doll's house, this too is a model of a complete building but it uses motifs and symbols rather than components (such as brickwork printed on paper and glued onto wall elevations). In this sense, the latter does not represent constructional information beyond a merely decorative level.

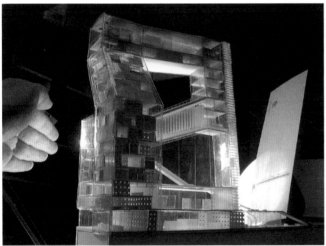

Above and above right

Ofis Arhitekti's competition design for Landmark Tower/U2 Studio in Dublin was conceived as a continuous programmatic strip, whose volume bends in three dimensions. The spatial complexity of the interior is revealed in the transparent surfaces of the model, as shown on the left. However, such descriptive models can have additional benefits and give an impression of how the final building will look on site,

as illustrated in the image to the right. The importance of a model in architectural competitions cannot be overstated, as it is perhaps the most effective means of communication to a judging panel in the absence of the design team.

Primarily, a descriptive model is built for the promotion of ideas, client consideration and public relations rather than decision making, since the architectural design is generally perceived to be determined by this stage. As such, the descriptive model is not conducive to significant alterations but is used to convey qualities of external form and internal relationships. A typical presentation model is a literal representation of the proposed work of architecture at a reduced scale. Within practical parameters, the closer the materials of the model approach those of the real building the better the model will be – since it describes the architecture best, i.e. the most precisely. This, in turn, makes the descriptive model a particularly useful tool for communicating a design to clients, members of the public, etc, who may not have the same ability as architects to perceive spatial information through drawings and other two-dimensional media. The finality of these descriptive models is perhaps best illustrated by their use. Beyond the realm of client presentations, such models of recent architectural designs – depicting both built and unbuilt works – may be found within transparent protective cases, usually set at eye level, almost in the manner of museum exhibits.

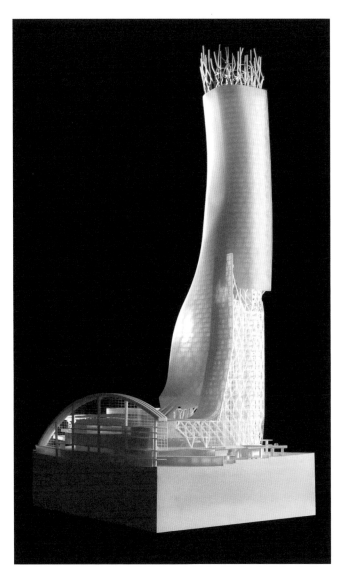

Opposite and this page
In this model for Morphosis'
design for the Phare Tower, Paris,
the translucent nature of the
modelling material embodies
the concept behind the project.
'Phare' is French for lighthouse,
as the external envelope enables
the structure to be seen beyond
it. The model accurately describes
the interplay between structure
and outer skin, whilst reinforcing
the luminous quality of the design.
This model has been extensively
published and exhibited, in
addition to being used to
communicate with the client.
Employed in this manner, the
power of such models to promote
a project affords designers a vital
method of engaging with the
public and the press.

Descriptive models may form a series in order to illustrate the progression of a design idea. Shown here are two design-development models (top row) alongside the final model (bottom left) of René van Zuuk's design for the Block 16 housing scheme in Almere, the Netherlands. When in use by the design team the development models' function is explorative, but when presented to external bodies at a later stage, their use is solely descriptive. The bottom right image shows the completed building.

It would be misleading, however, to imply that descriptive models are only applicable to client and public consultations. Exhibitions – particularly those including buildings that have long since perished, or major retrospectives of architects' work – will frequently showcase descriptive models specifically made for such events. These models are made either by the architect or a professional modelmaker under the architect's direction, except in the case of historical precedents. Here, they are more likely to be commissioned by an exhibition organizer, often as a more analytical version of the presentation model – their function, however, remains the same. Their purpose is explanatory, as they seek to describe architecture.

Below

For the competition to design the Guangzhou Sightseeing Tower, China, Rogers Stirk Harbour + Partners made a 1:500-scale model to describe their proposal. The level of precision and detail in such models is high, and this example is no exception. The central, structural element of this model was used to carry integral LED lighting to the vacuum-formed and scribed elliptical sphere containing a restaurant and public viewing area. The organic, tapering secondary structure is hand cut, formed and shaped from polyurethane resin block, while the base supports a vacuum-formed dome with a blue LED cluster spotlight up-lighting the whole structure.

Below right

This model was produced as part of an invited competition to design a skyscraper in Hamburg. Standing more than 3 metres (9 feet) high, the concept model showing a 'silhouette for a city like Hamburg' not only describes the design proposal but also becomes a sculpture in its own right. Representing, as it does, an unbuilt scheme, this model remains a testament to the earlier works of its designers, Coop Himmelb(l)au, and their referencing and sophisticated reconfiguration of Constructivism.

Bottom right

A descriptive model of Greg Lynn's design for the Ark of the World Museum and Visitors' Center, Costa Rica, provocatively conveys references to the jellyfish and brightly coloured indigenous tree frogs that inspired its form.

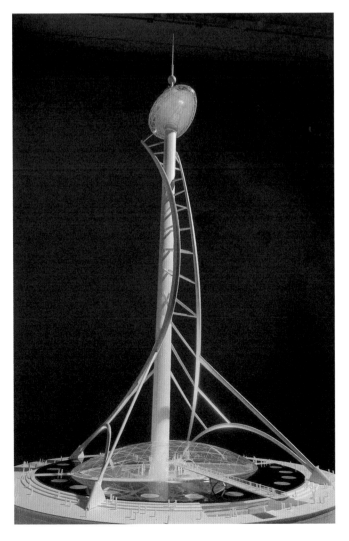

Predictive models

To appreciate the quantitative implications of constructing Steven Holl's design for the galleries in the Herning Centre for the Arts, Denmark, a full-size mock-up was produced. The curved roofs comprise a two-way truss system able to span in multiple directions, allowing for a freedom between the roof structure and the plan. The ceilings are of simple white plaster, complementing the geometry of the roof above. The roof would be anchored by introducing thin rods as tension elements within the clerestory glazing zone; these were intended to counterbalance any uneven forces over the gallery-wall support, like a see-saw tied down on each side. By making a predictive model in order to measure any potential difficulties on site, this example provided an effective method of testing the 'realities' of architecture.

The predictive model, by its very nature, is used to forecast the future. Predictive models are based on the assumption that the situation is an established rather than an emergent one. In the case of an established situation, the relevant environmental conditions are given and predictable, whereas in an emergent situation some of these conditions do not prevail and therefore the data generated is often qualitative rather than quantitative. This is a significant point, since an important distinction between predictive models and evaluative models is the type of data generated through their use. The function of predictive models is predominantly to produce quantitative data, whereas the use of evaluative models generally provides data of a qualitative nature. This is found to be particularly relevant to the case of architecture, a discipline in which predictive models enable the designer to measure the effects of changing design variables rather than perceive them, since there can be no 'difference which doesn't make a difference'. Another distinction between

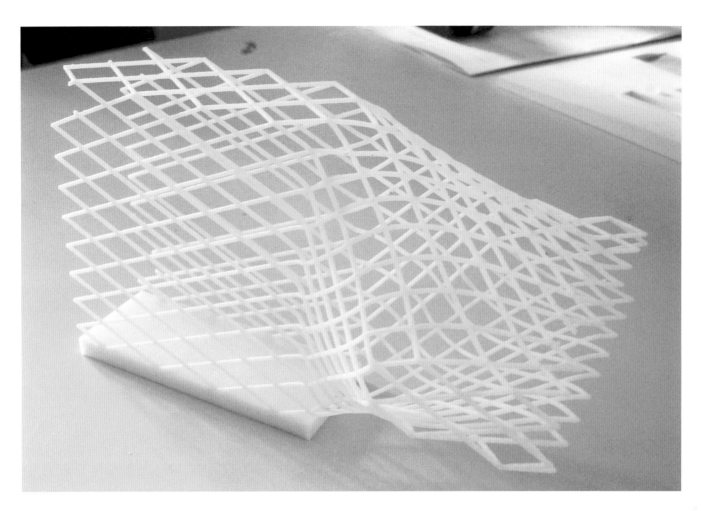

predictive and evaluative architectural models is seen
in the stage at which they are used during the design
process. As described above, an evaluative model is used
when many of the design variables have been decided
in order that their effects can be perceived. By contrast,
a predictive model is usually used earlier in the design
process so that such design variables can be determined
by its use.

Above
Predictive models can be produced
at a range of scales. In this case,
a complex three-dimensional net
structure was modelled in order to
enable the designer to predict the
areas of weakness.

Below
A 1:1-scale model of a seating
element made from high-density
foam, produced in order to predict
zones of structural deformation
through use prior to the
manufacture of the component on
a larger scale.

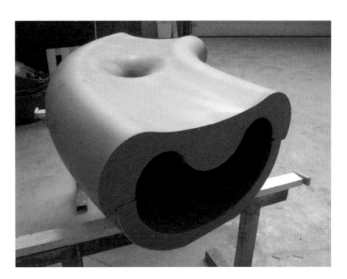

Case Study Detail models

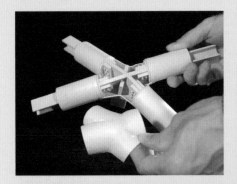
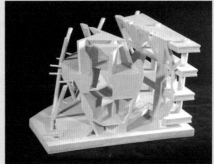
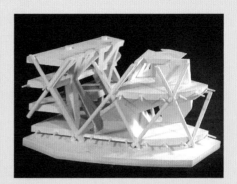

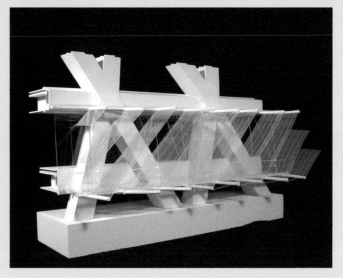

In architecture, predictive models are often used to allow us to forecast the impact of buildings within their environment. Most importantly, the function of predictive models is to produce quantitative data, i.e. those properties that can be measured. There are different types of architectural model that can be used to produce such data, for example: lighting models and wind-tunnel models. Lighting models are used to predict and measure the effect of varying combinations of natural and simulated electric lighting. This is particularly significant in light-sensitive spaces, for example those in museums and art galleries. However, an important distinction needs to be made here. Lighting models used as predictive tools are only concerned with measuring the relative light levels; they are not used to evaluate the quality of the lit environment, and therefore should not be confused with the specialized interior models built to investigate the impact of qualitative properties such as surface finish and colour. For the purpose of producing quantitative data, lighting 'models' usually comprise mathematical formulae rather than physical models. Wind-tunnel models may also be used to forecast any possible deformation in either the external envelope of a building or its individual construction components as a result of air pressure, suction and turbulence. This type of prediction is typically carried out using a specialized test chamber, in which smoke is added to the airflow to enable circulation patterns to be seen.

The complexity of the construction process involving innovative building forms and geometries frequently leads to the making of predictive models to allow designers to understand how the various components will connect together in three dimensions. This may lead to full-size prototyping as illustrated in the top left image on this page. However, in earlier stages of the design process, in which the exact structure may not yet be known, it may be useful to produce predictive models so that the various elements can be quantified prior to testing at a even more detailed level. The examples of 1:20 details shown here are from a wider series of models developed so that the changing effect of design variables could be measured.

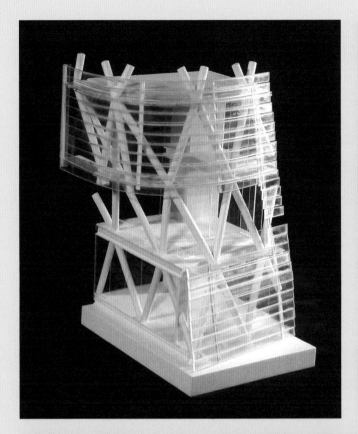

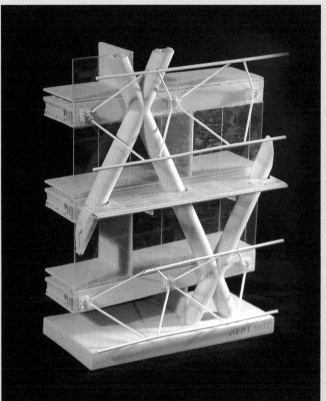

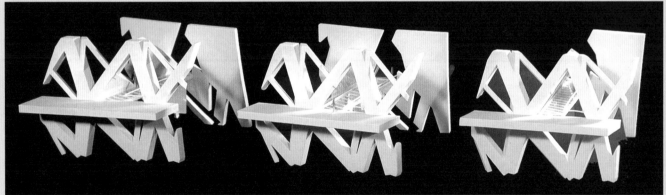

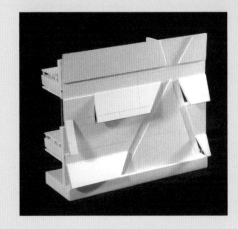

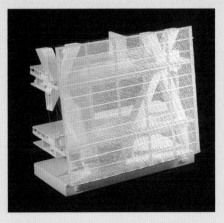

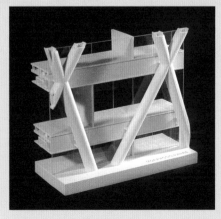

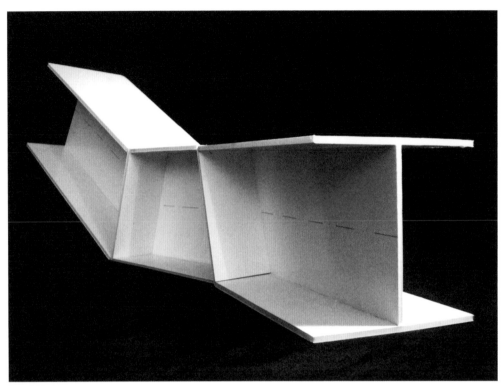

Left
The quantity of material required to make junctions and other building components can often be tested in modelling materials prior to production, which can involve more expensive and time-consuming fabrication processes. This predictive model of a structural element can be studied before it is made from steel.

Below
Predictive models enable communication between the design and construction teams to be effective, and facilitate further refinement of building details in order to avoid possibly costly difficulties on site. An inspection panel such as the one shown here may be built to allow all those concerned to check the implications of a design before the building is constructed.

Evaluative models

The purpose of the evaluative model is to explore or describe something such as properties or experiences that are not manifest in the model itself but are related to it. The evaluative model differs from predictive and explorative types since with these last-mentioned it is the model itself that attempts to assist the understanding of reality or a particular phenomenon. By contrast, the evaluative model seeks to assist such understanding by its use, and to do this it relies upon information and actions external to it. Not to be confused with predictive models, which produce quantitative data, evaluative models are intended to provide data of a qualitative nature, i.e. those properties whose variable effects can be perceived rather than measured. As a result, an evaluative model is typically – but not exclusively – used during the later stages of the overall design process, when many of the design variables have been determined. Using an architectural model for evaluative purposes is certainly not a new idea. The use of full-scale evaluative models for experiencing architectural components in situ is well documented throughout architectural history. It was particularly prominent in the Age of Enlightenment, when maquette makers held some superiority over practitioners of the more traditional technique of perspective drawing. The degree to which such models are detailed obviously depends upon the information to be evaluated, but the potential of representing all spatial, visual and tactile qualities at a level almost equal to that

of reality is naturally highly valuable. This application of models clearly has beneficial opportunities for the architect who wishes to articulate space and develop a design following the reaction of its intended users.

There is, of course, some overlap between these various model applications, depending on who is using them and why. This allows a transformation in the model type as a result of an exchange of intention, i.e. the evaluative model may also become a predictive model because it not only provides qualitative information but also data of a quantitative nature. This is particularly evident in the case of detailed and full-size mock-ups of building components and construction samples, since the form, materials and assembly have been determined and the model can now be used to predict structural behaviour – information that has to be measured rather than simply perceived – as well as evaluate it. Therefore, rather than explore or describe, for evaluative purposes, something that isn't manifest in the model itself but is related to it the model is now used to communicate the way reality is or could become, using the behaviour of the model itself.

A development model produced during the design process is photographed and montaged, to scale, onto its urban context. Techniques such as these permit the designer to evaluate the potential impact of ideas in terms of colour, material and form.

The qualities of light across form and within spaces are of primary concern to most architects. Through the making of a model at an appropriate scale, the design of even detailed elements can be interrogated and their characteristics perceived. In this example, the effects produced by a design for a solar shutter upon the domestic spaces behind can be evaluated. This is an important distinction, as an evaluative model communicates qualitative aspects rather than measurable ones – although this model could also have been used to take accurate light readings if required. By assessing the impact of the shutter on the interior, further decisions can be made about the pattern and size of apertures within the shutter itself.

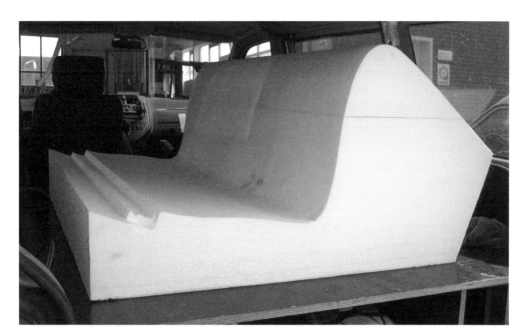

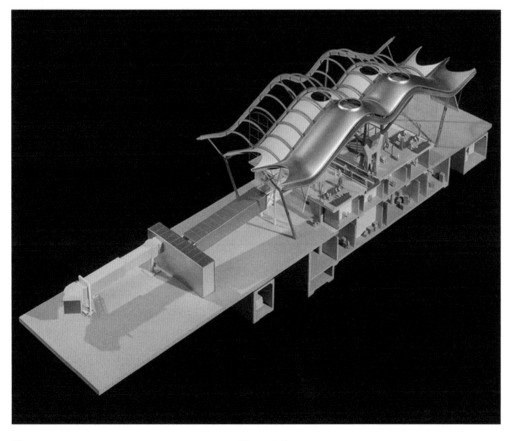

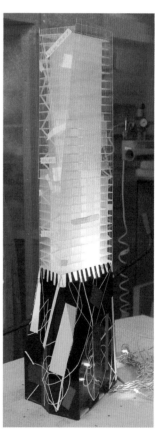

Left

A model of a seating design is shown here ready to be taken to site so that the designer can evaluate its physical impact. The use of evaluative models such as this enables designers to clearly demonstrate to clients, manufacturers and the public the scale and scope of their design ideas.

Above

This 1:200-scale sectional model of the design for Madrid's Barajas Airport by Rogers Stirk Harbour + Partners illustrates a cross-section through the satellite building within the air bridge. Each bay of the roof shows a different layer of construction, communicating the build-up to the final finish. This model was also used by the designers to evaluate the layout of the below-ground servicing and passenger-movement connections.

Above right

The application of evaluative models may vary, and it is not always for the qualitative appreciation of design elements. In this example, a design-development model has had lighting fitted into it prior to a presentation so that the design team can observe the effects this has within the model. This highlights the dynamic aspect of models and their ability to be appropriated for different functions by different users.

Case study Evaluating a space

The pursuit of excellent spatial qualities within a design is at the core of the architectural profession. For designers, it is important that creative ideas are experimented with in order to ensure that as many design opportunities as possible are investigated. Evaluating the full implications of a design is not always achievable, for instance, under typical daylight conditions. Using a model made to represent an events space within a public building, the designer tried various lighting effects using different light sources and colours. The results were captured using a long exposure time on the camera, enabling lights to be moved around and within the model. This allows the space to be perceived as it may appear during a performance, and allows the potential atmosphere and mood of the design to be communicated to others – information which cannot be gained from viewing the model in isolation.

STEP BY STEP EVALUATING THE EFFECTS OF A FAÇADE

A single detail may embody the concept of a building, and it is this level of consideration across a range of scales that contributes to good architectural design. The making of detail models is often a useful exercise, as they are effective tools for understanding how components may be assembled together in three dimensions and the resultant effects can then be observed. This evaluative model was produced so that the qualities of an etched glazing façade could be studied by the design team and subsequently presented to the client.

1 An initial visualization drawing is produced using digital software.

2 The designers then make a 1:20-scale sectional model of the fritted glass panel using etched acrylic sheet.

3 The effects of different coloured lighting are evaluated using a component model, affording further refinement of design decisions to be made.

Above

This 1:50-scale white-and-clear-
acrylic model, with vacuum-
formed roofs and laser-etched
façade, was used as a working
tool for the design team fully to
evaluate the proposal in three
dimensions and to test out various
design elements. It has since
become a presentation model in
the client's existing headquarters
and also featured in a retrospective
exhibition of the designer's work,
thereby shifting from an evaluative
application to a descriptive one.

Right

This 1:100-scale model of a façade
was quickly produced and used
by the designers to help them
understand the depth of façade
elements, the play of shadow
and the use of colour on the
balconies. Lighting from within
adds further depth to the model
in photographs, and facilitates
greater evaluation of the
design's features.

Explorative models

The main purpose of the explorative model is to discover other realities by speculation. This speculative process involves systematically varying the parameters used in the descriptive model in order to identify those alternatives that are logically possible. In the context of architecture, models produced as part of the design-development process can be considered explorative models. Explorative models, along with drawings, can be used as a method of refining judgements, making decisions or conveying information – factors that are at the very core of architectural design. Explorative models, by their very nature, are used to try out and test different ideas at different degrees of scrutiny, and consequently they are used at various stages during the design process. However, they are typically employed throughout the early stages of design development, when the designer's ideas are at their most novel, and then subsequently developed prior to specific properties of the design being investigated through the application of other types of models, as described earlier in this section. The key characteristic of an explorative model is that it is concerned with testing out new ideas. These may not be exclusively structural concepts, but may often involve exploring different shapes, geometries or construction methods that will emerge from experimenting with form and material in different ways.

An explorative model used by MUF as part of a feasibility study for its Museum of Women's Art, applied as a conceptual design model that both tested and refined the client's ambition. The study questioned the paradox of revealing a hidden canon of work only to enclose it again in another hermetic institution. As such, the model explores strategies and spatial arrangements for making connections and juxtapositions, both actual and visual, that allow for the effective curating of the museum.

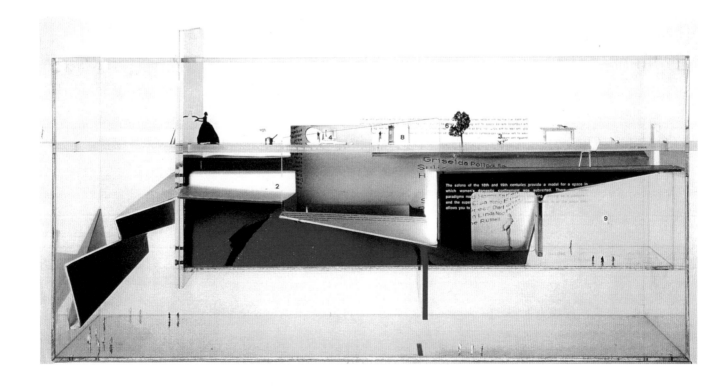

Explorative models may cover a vast spectrum of methods for representing ideas. From the highly detailed to the purely diagrammatic, they can communicate the subtle characteristics of a design's underlying principle – or even act as a symbolic object for the project.

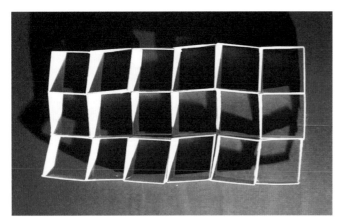

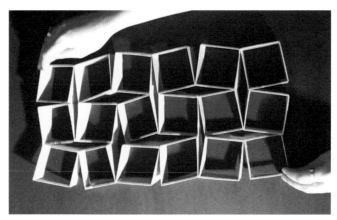

Above images
UNStudio's design for Ponte Parodi, Genoa, was originally conceived as a three-dimensional plaza. This series of images demonstrates how a model was used to explore this concept and develop its low-slung, undulating outlines.

Below left and right
Once generated through initial models, the design was further refined and communicated through a series of explorative models made from different materials.

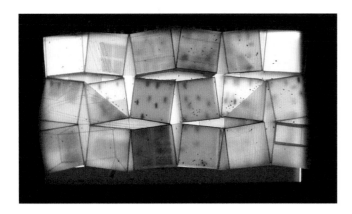

In direct contrast to the descriptive model described earlier, the design development model affords quick modification – or even radical change – in order to test ideas, rather than preciousness. That is not to say that there can be no refinement in such models. However, rather than the focus of a design development model being on the quality of the finish it is on the refining of the emergent design concept. Consequently, many explorative models in architecture are quickly produced, and may appear unfinished since they function as three-dimensional sketches to help develop the design. Such models, however crudely built, give expression to ideas, and help a designer explore what is possible and communicate new design notions. As has already been illustrated within this book so far, the design development process encompasses a range of models from basic conceptual forms to detailed component inquiries. Such models are typically produced rapidly and inventively, using a variety of mixed media to symbolize, for example, relationships between the components of a building concept or its response to its context.

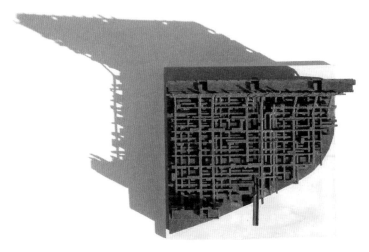

Below

This explorative model was used to develop a façade design. Made from mirror-faced cardboard, initial ideas about geometry and facets could be pursued by the designer before its development in detail.

Above

Morphosis' project for social housing in Madrid is based on the notion of porosity. The application of an explorative model to develop this concept as a reflection of the social ideals within the scheme, also allowed the designers to investigate and refine its intricate folded-lattice structure. (For more images of this project see page 137.)

Below

A highly sculptural model produced to express a designer's initial reactions to the urban context of a design brief. The flows of paper explore the emotional and physical areas of compression and openness across the site.

Right

The model shown in this trio of images was built to allow the designer to explore ideas for a building as a linear series of programmed spaces, which are then folded and stacked on top of each other owing to site conditions.

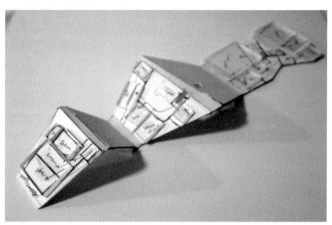

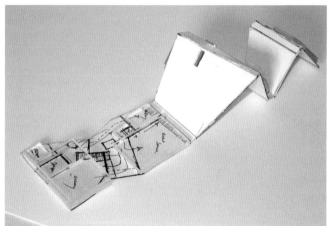

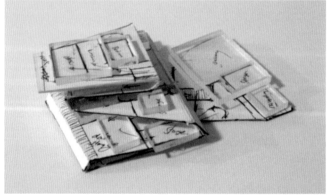

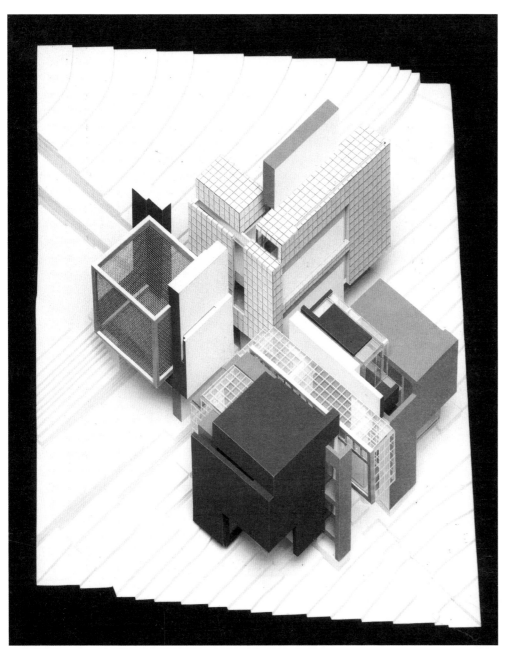

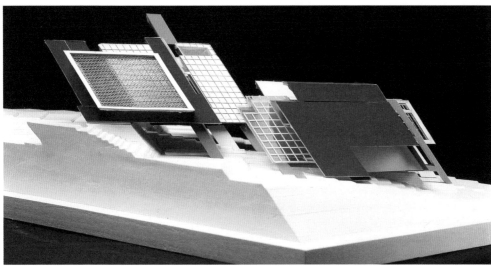

Peter Eisenman's exhaustive inquiries into geometry have led to an extensive catalogue of explorative models that document his various design processes. In this example, the design for House X was initially developed as a juxtaposition of four squares whose structure is incomplete. The dexterity with which such an understanding of space is handled and questioned is manifest through the production of an axonometric model exploring the design. Compare the shot of the model shown above, taken from the viewpoint of a single-point perspective, with the image below, which reveals a completely different aspect of the design.

Below
The exploration of a creative idea is rarely a single-stage process, and often requires much iteration to be carried out as the design is revised and enhanced. These examples of a street-lighting element document the designer actively engaged in the quest for the most suitable solution.

Opposite
Design development plays a crucial role in the practice of architecture, as ideas are extensively explored, altered or relegated to the conceptual scrapheap. Models may provide effective expression of a designer's thoughts and how they develop over time, as illustrated by this sequence of models detailing the evolution of an entire building form.

Designers who prefer to develop new spatial ideas and formal inquiries directly in three dimensions use explorative models as an initial design tool. Whilst commonly perceived as a personal and embryonic three-dimensional sketch, the spontaneity and immediacy of such models is of particular significance since it enables them to be a very flexible medium. Being quickly produced from easily worked materials, it is their singular focus on contrast in shape, size and colour that facilitates rapid change and development. In addition, whereas the presentation model seeks to provide a holistic view of the finished architecture, the design development model may be produced in order to explore particular components of the design. As the form of a new architecture emerges, a whole series may be constructed of more specialized models that respond to questions arising from the initial evolution of that architectural form. These are study models, i.e. complete or part models specifically built to address certain issues. By working directly in space, albeit at small scale, concepts are formed and refined as a result of their exploration in three dimensions – a process in which options remain open in design routes, which might not be readily apparent to the designer using two-dimensional drawing methods alone.

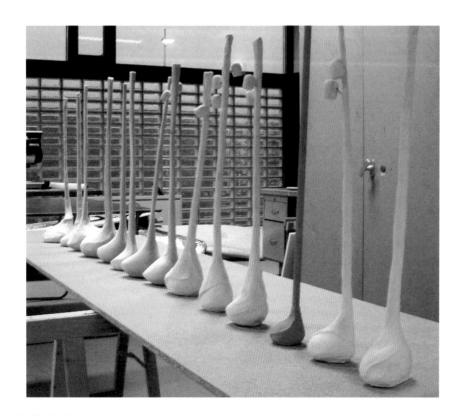

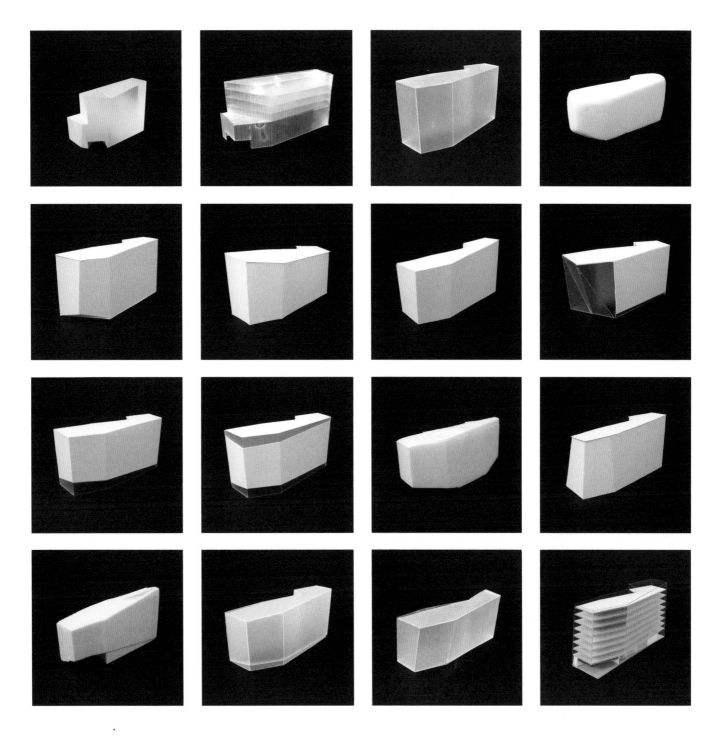

Whilst there are many different types of architectural model, the use of the explorative model can be a particularly important part of the design process. Although the majority of architectural practices commission or build the more familiar descriptive or presentation model – immaculately crafted and expensively built – for client and/or public consumption, many still use explorative models as tools for engaging with the spatial problems and other aspects of a design. Perhaps, however, we are currently witnessing a renaissance, in which the explorative model is used and shared with client and public groups. Once the

hidden tool of the architectural designer, these models are increasingly being applied as discursive vehicles in communication with those outside of the immediate design team – an attitude expressed by Rem Koolhaas, who refers to this process as showing clients creative thinking produced in its raw form.

STEP BY STEP EXPLORING MATERIALS THROUGH MODELS

Explorative models provide exciting vehicles, with which to experiment with materials and spatial ideas before a design becomes defined. Try working with as many different materials as possible in order to expand both knowledge and understanding of them, since this may prove highly useful in future design projects and will engage the imagination.

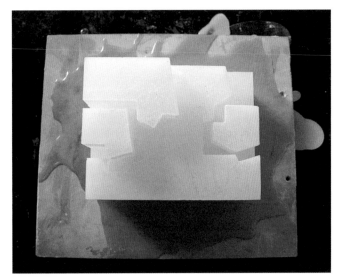

1 A wax positive model is stuck to a board with some additional melted wax.

2 A liquid 'slip' is applied in order to allow the wooden boards, which will be placed around the sides to form a box, to be easily removed.

3 Plaster is poured into the wooden box and allowed to dry out completely.

4 The wax is infused into the plaster.

5 The wax is then burnt out.

6 A negative cast of spaces is revealed where the wax had been.

7 These spaces can then be explored architecturally.

8 The simple addition of various scales of figures allows the same spaces to be interpreted differently.

Case study Evolving Scars, Bernard Khoury

The intention of a model may be to explore theoretical ideas and question preconceptions about architecture and spatial characteristics. For example, the Evolving Scars project by Bernard Khoury sought to address the rehabilitation and reconstruction of the Beirut Central District, and proposed turning the process of demolition of war-damaged buildings in the city into a collective architectural experiment. The project consists of a temporary transparent skin that is implemented around the perimeter of a ruin, and a 'memory collector' that deploys itself within the ruin while collecting data. The intensity of collecting information is translated by the gradual demolition of the existing edifice. The 'remains' of the ruin are collected and contained within the transparent peripheral membrane. The process ends with the complete demolition of the ruin, and is an attempt to transform the demolition of buildings into an architectural act – albeit an ephemeral one.

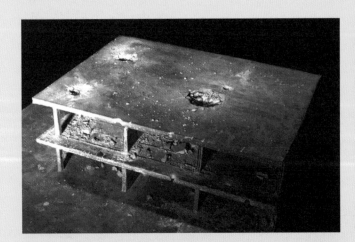

Modelling the future

This book has described a vast spectrum of models, investigating what they may be made from, explaining why they may be produced and discussing how they provide different modes of inquiry in architecture. The use of models as a medium and mechanism through which the development of architectural knowledge and design innovation can be produced remains paramount to the discipline, and this appears very unlikely to change. However, what is less certain is how such practice will be transformed as a result of future developments and which aspects may evolve while others recede. From the current situation it appears that the role of models as generative and representative tools in architecture will continue to flourish as the productive differences between various media and techniques continue to grow. Perhaps the most exciting field lies in the increasing overlaps between digital and traditional design processes and techniques. Richard Armiger of Network Modelmakers speculates that there may be a number of evolutionary strands in architectural modelmaking, but currently the most engaging is that 'as technology progresses the limitations on durability of some techniques, such as rapid prototyping, will be overcome and then integration with more traditional methods will be really interesting.' (Interview with author, 11 September 2008.)

The emergence of new computational modelling software that enables parametric systems and complex biological organizations to be generated and explored in design terms has begun to offer avenues for the modelmaker that had not previously existed. Such development is characterized by the contemporary position offered by Bob Sheil: 'Never before have there been so many, or such varied, techniques and methods at our disposal, each with the capacity to leap only previously imagined frontiers. Designing has become a liquid discipline pouring into domains that for centuries have been the sole possession of others, such as mathematicians, neurologists, geneticists, artists and manufacturers. Post-digital designers more often design by manipulation than by determinism, and what is designed has become more curious, intuitive, speculative and experimental.' (Sheil, B., ed., *Protoarchitecture: Analogue and Digital Hybrids*, London: John Wiley & Sons, 2008, p.7). These huge transformations in design processes have implications far beyond the discipline of architecture as more and more research and developments are being conducted at cross-disciplinary levels around the globe. The surge of interest in this field is perhaps typified by major exhibitions exploring the design possibilities opened up by important advances in technology, for example 'Design and the Elastic Mind', held at MoMA, New York, in 2008. To conclude, the significance of the model as a medium for the advancement of architectural knowledge and as a dynamic tool that can catalyse and renegotiate relationships between concepts, techniques and different modes of inquiry cannot be overstated. The realms of design opportunities await further exploration.

The Summer Show at the Bartlett School of Architecture, University College London, in 2009, reinforces the importance of the role of the model as a design and communication tool.

Right and centre right

The generative possibilities of emergent computer software provide a rich area of experimentation for designers and enable complex geometries to be readily incorporated in modelmaking. These two examples demonstrate the integration of digital techniques to explore new design ideas in relation to structure, in this case using laser-cut card (top) or Rapid Prototyping (below).

Below

Through the integration of concepts, techniques and modes of inquiry, the future of modelmaking and the production of protoypes that develop new knowledge and discourse in architecture is reaching new territories, as exemplified by the summer pavilion at the Architectural Association, London, 2009.

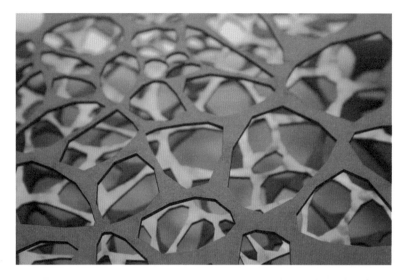

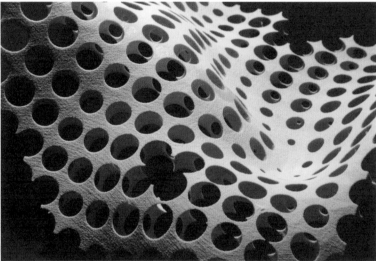

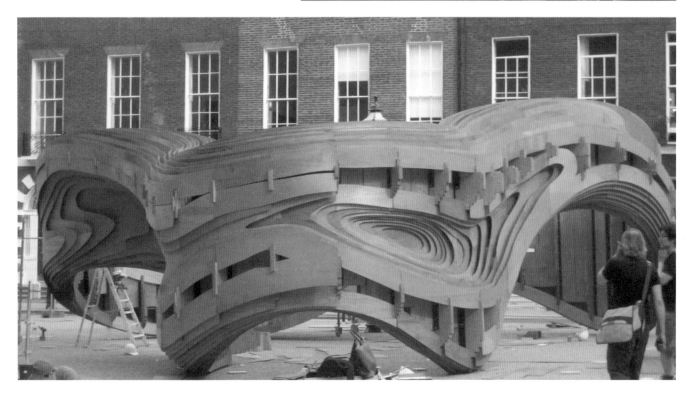

Glossary

There are a number of key words relating to the practice of architectural modelmaking that are typically understood by everyone but may be interpreted differently, giving rise to ambiguity if not misunderstanding. In the interest of clarity, it is valuable to provide a definition of the more significant words and phrases used in this book. Perhaps most importantly, it is worth starting with a term that can result in much confusion depending on how it is used: model.

Model

There are multiple entries under the dictionary definition of a model although 'a representation, generally in miniature, to show the construction or appearance of something' may seem the most appropriate in this context. However, even this definition is rather limiting as it does not emphasize the dynamic role the model has. For the purpose of this book it is important to consider a model as both the medium and mechanism through which design ideas are generated and represented.

CAD Computer-Aided Design (or Drafting) software is used by architects and students to develop and communicate their design ideas. Different software packages can produce different results: some only produce two-dimensional drawings while others are capable of sophisticated three-dimensional renders or animations.

CAD/CAM This term is a shortening of Computer-Aided Design (CAD) and Computer-Aided Manufacturing (CAM). CAD/CAM software uses the information from CAD drawing tools to describe geometries used in the CAM portion of the programme to define a toolpath that will direct the motion of a machine tool to machine the exact shape that was drawn.

CNC is an acronym for Computer Numerically Controlled equipment, which is programmed and controlled by computer. The advantages are that once the CAD data has been drawn, this type of machine can offer very short set-up times and the flexibility to run batches ranging from one-offs to large numbers of repetitive components.

Concept is an idea of something formed by mentally combining its features as a construct. A concept is often the driving force behind an architectural design because it initiates the design process and threads through its development.

Generative refers to the capacity of a model to be able to create or produce design ideas rather than simply represent them.

Photomontage is a technique that combines one image of a design proposal with another. Carefully taken photographs of physical models can produce even more impressive results when merged with contextual information in digital photographs or images.

Prototype is an original type, form, or instance of something that serves as a typical example, basis, or standard for further development. In architecture, the experimental nature of the discipline often requires the need for prototyping to enable the designer to explore, test and evaluate potential solutions.

Rapid prototyping is an automatic process of making physical models using additive manufacturing technology by taking the data from CAD files and converting it into successive layers of liquid, powder, or sheet material, to build up a model from a series of cross sections. These layers are then bonded together to produce the final form. The key advantage of this technique is its ability to create almost any form or geometric feature.

Representation carries several different meanings but in this context it is used to describe the communicative aspects of a model, that is to say it is a mode of expression for design ideas rather than a generator of them.

Scale is described as a ratio in relation to reality and enables architects and students to accurately describe design ideas in drawings and models. The specific scale used may be metric or imperial but care should always be taken when converting from one scale to another.

Section, in relation to drawing, refers to a vertical cut made through a space or building. For modelmaking purposes this convention can allow the designer to explore connections between the interior and exterior of the design, as well as internal relationships.

Further reading

Abruzzo, E., Ellingsen, E. and Solomon, J. D., *Models*, 306090, volume 11, Princeton Architectural Press, New York, 2007

Allen, S. and Agrest, D., *Practice: Architecture, Technique and Representation*, G&B Arts International, Amsterdam, 2000

Antonelli, P., ed., *Design and the Elastic Mind*, The Museum of Modern Art, New York, 2008

Beckmann, J., ed., *The Virtual Dimension: Architecture, Representation and Crash Culture*, Princeton Architectural Press, New York, 1998

Busch, A., *The Art of the Architectural Model*, Design Press, New York, 1991

Callicott, N., *Computer-aided Manufacture in Architecture*, Architectural Press, Oxford, 2001

Dernie, D., *Architectural Drawing*, Laurence King Publishing, London, 2010

Dunn, N., *The Ecology of the Architectural Model*, Peter Lang, Oxford, 2007

Eissen, K., *Presenting Architectural Designs*, Phaidon Press, London, 1990

Frampton, K. and Kolbowski, S., eds, *Idea as Model*, catalogue 3, Institute for Architecture and Urban Studies, Rizzoli International Publications, New York,1983

Frascari, M., Hale, J., and Starkey, B., eds, *From Models to Drawings: Imagination and Representation in Architecture*, Routledge, New York, 2007

Healy, P., *The Model and its Architecture*, 010 Publishers, Rotterdam, 2008

Hohauser, S., *Architectural and Interior Models*, Van Nostrand Reinhold, New York, 1984

Janke, R., *Architectural Models*, Thames & Hudson, London, 1968

Knoll, W. and Hechinger, M., *Architectural Models: Construction Techniques*, J. Ross Publishing, Fort Lauderdale, 2007

Leach, N., ed., *Rethinking Architecture: Reader in Cultural Theory*, Routledge, New York, 1997

Mills, C., *Designing with Models: A Studio Guide to Making and Using Architectural Design Models*, John Wiley & Sons, Inc., New York, 2000

Moon, K., *Modeling Messages: the Architect and the Model*, Monacelli Press, New York, 2005

Morris, M., *Models: Architecture and the Miniature*, Wiley-Academy, Chichester, 2006

Oswald, A., *Architectural Models*, DOM Publishers, Berlin, 2008

Porter, T., *The Architect's Eye*, E & FN Spon, London, 1997

Porter, T. & Neale, J., *Architectural Supermodels*, Architectural Press, Oxford, 2000

Ratensky, A., *Drawing and Modelmaking: A Guide for Students of Architecture and Design*, Whitney Library of Design, New York, 1983

Sheil, B., ed., *Design through Making*, John Wiley & Sons, London, 2005

Sheil, B., ed., *Protoarchitecture: Analogue and Digital Hybrids*, John Wiley & Sons, London, 2008

Smith, A. C., *Architectural Model as Machine*, Architectural Press, Oxford, 2004

Sutherland, M., *Modelmaking: A Basic Guide*, W. W. Norton & Co., New York, 1999

Vyzoviti, S., *Folding Architecture: Spatial, Structural and Organisational Diagrams*, BIS Publishers, Amsterdam, 2003

Index

Page numbers in *italics* refer to
picture captions

Picture credits

Front cover **Feilden Clegg Bradley**
Back cover **Rogers Stirk Harbour + Partners**
1 **Ofis Arhitekti**
3 **Eduardo Souto Moura**
4 **Grafton Architects**
6 **Office for Metropolitan Architecture / ©DACS 2010**
7 left **Rogers Stirk Harbour + Partners**
7 right **Antoine Predock**
8 left **6a Architects**
8 right **Alsop Architects**
9 **Grafton Architects**
10 **UNStudio**
11 top **COOP HIMMELB(L)AU / Markus Pillhofer**
11 bottom **Alsop Architects**
12 left **Rogers Stirk Harbour + Partners**
12 right **Alsop Architects**
13 top **Daniel Libeskind**
13 bottom **COOP HIMMELB(L)AU / Markus Pillhofer**
14 left **Alinari Archives, Florence**
14 right **RIBA Library Photographs Collection**
15 **© National Maritime Museum, Greenwich, London**
16 left **RIBA Library Photographs Collection**
16 right **Courtesy the Catherine Cooke Collection, University of Cambridge**
17 top **©National Museums Liverpool, Walker Art Gallery**
17 bottom **RIBA Library Photographs Collection**
18 **RIBA Library Photographs Collection**
19 top **RIBA Library Photographs Collection**
19 bottom **Rudi Meisel/Visum**
20 top **Joseph Haire**
20 bottom **Tarek Shamma**
21 **Simon Rodwell**
23 1. **Nick Dunn**, 2. **Katherine Burdett**, 3. **Nick Dunn**, 4. **Nick Dunn**, 5. **Katerina Scoufaridou**
25 **Nick Dunn**
26 **Katherine Burdett**
28 **Grafton Architects**
29 top **Antoine Predock**
29 bottom **Tim Hursley**
30 **Ofis Arhitekti**
31 **Holodeck.at**
32–33 **The AOC**
34 top and centre **Grafton Architects**
34 bottom **Richard Brook**
35 left **Eduardo Souto Moura**
35 bottom **Thomas Hanson**
36 **Alsop Architects**
37 top row and left **UN Studio**

37 bottom right **Christian Richters**
38 top and bottom left **Steven Holl Architects**
38 bottom right courtesy **Steven Holl Architects/Iwan Baan**
39 top **Richard Brook**
39 centre **COOP HIMMELB(L)AU / Markus Pillhofer**
39 bottom **Delugan Meissl Associated Architects**
40 **Holodeck.at**
41 **Ofis Arhitekti**
42 centre **Alison Brooks Architects**
42 bottom **6a Architects**
43 top **Daniel Libeskind**
43 centre **Daniel Libeskind / Torsten Seidel**
43 bottom **with permission of the Royal Ontario Museum ©ROM**
44 **Grafton Architects**
45 top **MacKay-Lyons Sweetapple**
45 bottom **Dixon Jones**
46 top **Meadowcroft Griffin**
46 bottom **Grafton Architects**
47 **6a Architects**
48 top **6a Architects**
48 bottom **MacKay-Lyons Sweetapple**
49 **Grafton Architects**
51 **team-bau**
52 top **Katerina Scoufaridou**
52 centre left **René van Zuuk**
52 centre right top and bottom **Greg Lynn FORM**
52 bottom **Nick Dunn**
53 **Daniel Libeskind / L. Haarman**
54 top **Office for Metropolitan Architecture / ©DACS 2010**
54 bottom left **Rogers Stirk Harbour + Partners**
54 bottom right **Archi-Tectonics**
55 top **Katherine Burdett**
55 bottom **Alsop Architects**
56 **COOP HIMMELB(L)AU / Markus Pillhofer**
57 top **COOP HIMMELB(L)AU/ ©Gerald Zugmann**
57 bottom **COOP HIMMELB(L)AU/ ©Gerald Zugmann/ www.zugmann.com**
58 **Rogers Stirk Harbour + Partners**
59 top **Steven Holl Architects**
59 bottom **Alsop Architects**
60 top **Nick Dunn**
60 bottom **team-bau**
61 top **Katerina Scoufaridou**
61 bottom left **ALA**
61 bottom right **Nick Dunn**
62 **Office for Metropolitan Architecture / ©DACS 2010**
63 top **Nick Dunn**

Author's acknowledgements

I am grateful to everyone who contributed to this book, in particular to David Dernie whose support and collaboration initiated it; to Richard Armiger of Network Modelmakers, whose endless enthusiasm and participation provided valuable insights; to Gemma Barton, Yvonne Baum and Barry McCall, who gave considerable assistance in the early stages of collecting material from practitioners; to Jim Backhouse, whose experience, generosity and participation always proved to be highly engaging; to Katherine Burdett, who donated a significant amount of time and material; to the staff and students of Manchester School of Architecture for providing creative comments and innovative examples respectively; to Philip Cooper and Liz Faber of Laurence King for their tireless advice and patience throughout the production of this publication; and to my family and friends whose support enabled it to be written.